9th EPS-QEOD Europhoton Conference on Solid-State, Fibre, and Waveguide Coherent Light Sources (EUROPHOTON 2020)

AA004505

EPJ Web of Conferences Volume 243 (2020)

Online
30 August - 4 September 2020

Editor:

P. Helfenstein

ISBN: 978-1-7138-1921-9

Printed from e-media with permission by:

Curran Associates, Inc.
57 Morehouse Lane
Red Hook, NY 12571

Some format issues inherent in the e-media version may also appear in this print version.

No changes have been made to the content of these proceedings. There may be changes to pagination and minor adjustments for aesthetics.

Printed with permission by Curran Associates, Inc. (2021)

For additional information, please contact EDP Sciences – Web of Conferences
at the address below.

EDP Sciences – Web of Conferences
17, Avenue du Hoggar
Parc d'Activité de Courtabœuf
BP 112
F-91944 Les Ulis Cedex A
France

Phone: +33 (0) 1 69 18 75 75
Fax: +33 (0) 1 69 28 84 91

contact-edps@webofconferences.org

Additional copies of this publication are available from:

Curran Associates, Inc.
57 Morehouse Lane
Red Hook, NY 12571 USA
Phone: 845-758-0400
Fax: 845-758-2633
Email: curran@proceedings.com
Web: www.proceedings.com

TABLE OF CONTENTS

edp sciences

EPJ Web of Conferences

Statement of Peer review

In submitting conference proceedings to EPJ Web of Conferences, I certify to the Publisher that I adhere to the **Policy on Publishing Integrity** of the journal in order to safeguard good scientific practice in publishing.

1. All articles have been subjected to peer review administered by the proceedings editors.
2. Reviews have been conducted by expert referees, who have been requested to provide unbiased and constructive comments aimed, whenever possible, at improving the work.
3. Proceedings editors have taken all reasonable steps to ensure the quality of the materials they publish and their decision to accept or reject a paper for publication has been based only on the merits of the work and the relevance to the journal.

Title, date and place of the conference EUROPHOTON 2020 ON SOLID-STATE, FIBRE AND WAVEGUIDE COHERENT LIGHT SOURCES. 30 AUGUST – 4 SEPTEMBER 2020

Proceedings editor(s): EUROPEAN PHYSICAL SOCIETY

EUROPEAN PHYSICAL SOCIETY
6 rue des Frères Lumière
68200 MULHOUSE
Tél 03 89 32 94 40

Date and editor's signature 06.10.20

EPJ Web of Conferences **243**, 01001 (2020) https://doi.org/10.1051/epjconf/202024301001
EUROPHOTON 2020

Power Scaling of Lasers

R. Paschotta
RP Photonics Consulting GmbH, Waldstr. 17, 78073 Bad Dürrheim, Germany

In the context of high power laser research and development, the term "power scaling" is very often used. However, very frequently this is done without using a clearly defined concept, although such concepts are quite common in other areas of technology such as computing and manufacturing. Correspondingly, the terms "scalability" and "power scaling" are often used in a vague or even meaningless manner. Clarifying such issues can create a lot of important insight.

This course explains power scalability of laser architectures as a precisely defined concept, and demonstrates with a number of examples that this is very useful in particular for evaluating future potentials of various kinds of high power lasers and thus for guiding further research and development. First, sensible criteria for a well-defined scaling procedure are discussed, the existence of which lies at the heart of any true scalability. Such a power scaling procedure describes how to transform an existing and working laser design into another design with e.g. two or ten times the output power. The essential features which must be demanded from a true scaling procedure are the following:

- The procedure must be systematic and well defined. For example, the form of such a procedure could be: For doubling the output power, double the pump power, apply it to twice the mode area in the gain medium, reduce parameter X by one half while keeping parameter Y constant.
- The procedure must be repeatable, allowing to vary the output power in a large range. This requirement has several important implications. In particular, application of the procedure should not spoil essential features of the laser, such as e.g. its power conversion efficiency or beam quality. Also, the procedure must be devised so that enhanced laser designs do not rely on increasingly critical specifications for the components to be used in the laser. Finally, the procedure must not make one or more central challenges, such as e.g. the handling of high optical intensities or strong thermal effects, significantly more severe as the power is increased.

Any laser architecture involves a large number of technical aspects, the scaling properties of which can be separately analyzed. Examples for such aspects are the involved optical intensities, thermal effects, or required pump brightness. Even though scalability of a laser architecture as a whole requires scalability of all such aspects, the study of scaling properties of isolated aspects can produce important insight for essentially any kind of laser. The analysis of current technical obstacles is only a starting point. It is vital to precisely understand (a) the physical details of limiting effects, (b) their impact on the further development (not only how strong they are, but also "how they scale"), and (c) available options to mitigate or eliminate those effects of strongest impact without introducing or exacerbating other problems. For issues (b) and (c), it is essential to recognize the existence or non-existence of a scaling procedure which takes into account the potentially limiting effects.

Plenty of examples can illustrate the usefulness of such thoughts. There are cases where a certain effect is certainly not weak even in low-power laser designs, but nevertheless does not prevent scaling to much higher powers, because a scaling procedure exists which allows to keep the critical parameters under control even at highest powers. A good example for this is the finding that thermal effects in saturable absorbers for passive mode locking, which can be significant even in low power lasers, have not prevented the power scaling over orders of magnitude; not recognizing certain scaling properties was the mistake which previously made many researchers believe that semiconductor absorbers would be unsuitable for high powers, at least with known materials and designs. On the other hand, effects which are negligible in low-power devices can later constitute limits of power scaling. For example, dispersion compensation can be a severe challenge in mode-locked high power lasers, while high power fiber lasers and amplifiers with diffraction-limited output will soon hit limits associated with nonlinearities and optical damage. Another interesting observation is that the power scaling properties of saturable absorbers can be very different for laser devices with high pulse repetition rates, creating a need for substantially revised designs even for much lower power levels than are easily achieved with lower pulse repetition rates.

Such examples convincingly demonstrate that a well-founded analysis of scaling properties is vital for understanding the impact of particular technical problems, assessing the value of proposed countermeasures, and judging the future potentials of different technological approaches. In this way, work on high power lasers can profit from the introduction of clearly defined terms and sensible criteria, and from the concrete analysis of important scaling properties. As a final remark, variations of the concept can be similarly useful for studying the scaling of other quantities than output power, such as e.g. pulse repetition rates of mode-locked or Q-switched lasers.

References

[1] R. Paschotta, "Power scalability as a precise concept for the evaluation of laser architectures", arXiv:0711.3987v1, http://www.arxiv.org/abs/0711.3987 (2007)

Double-clad fibers for high-power fiber lasers

Pavel Peterka
Institute of Photonics and Electronics, The Czech Academy of Sciences, Chaberska 57, 182 51 Prague, Czechia

Fiber lasers are considered as one of the youngest and most rapidly developing branch of lasers. In fact, we may observe recurrent increased interest in fiber lasers in time driven by various motivations and societal conditions. Although fiber lasers were proposed in 1960, even before the first laser light flash and before the first single-mode optical fiber was drawn, the fiber laser devices became hot-topic of research only in mid-80's as the invention of erbium-doped fiber amplifiers revolutionized the world of telecommunications and fueled the global boom of the internet in 90's. A second revolution started just at the beginning of the third millennium. With the availability of high-power, high brightness diode lasers to pump rare-earth-doped double-clad (DC) fibers, it began the race for high power from Yb-doped single mode fiber lasers at wavelength at around 1 μm [1, 2]. Major breakthroughs involve development of various kinds of components and fiber design for efficient coupling and combination of the optical pump and laser signal into the DC active fiber, microstructure and air-clad designs of the DC fiber and namely the invention of large mode area (LMA) fibers together with methods to maintain single mode operation of LMA fibers.

The high-power operation of fiber lasers was enabled by the invention of cladding pumping within a DC fiber structure. Such a fiber serves as an efficient transformer of the low-brightness, high-power radiation of the laser diodes (coupled into the large area inner cladding of the DC fiber) into a high-brightness, high-power laser beam coming out from the rare-earth-doped, narrow fiber core. Since the most common circular shape of optical fibers provides poor effective absorption of the pump, various cross-sectional shapes of DC fibers have been investigated both experimentally and theoretically in order to enhance the absorption of the multimode-pump, see examples in Fig. 1(a) [3-5]. The beneficial effect of mode mixing of the pump radiation by unconventional coiling was also observed, see example of the twisting effect of the two-fiber bundle in Fig. 1(b) [3].

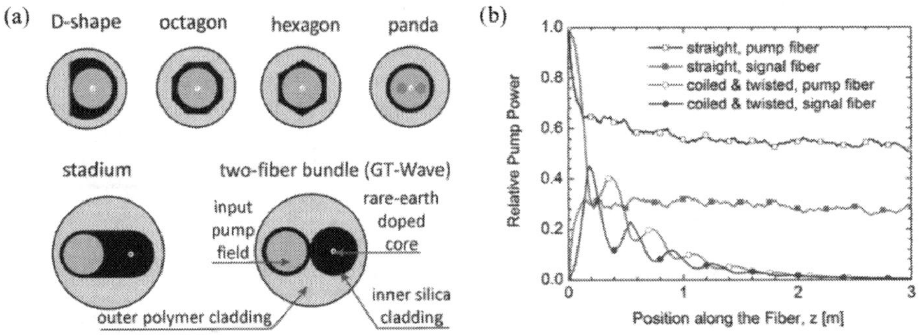

Fig. 1 (a) Examples of double-clad fiber cross sections of the inner cladding with broken circular symmetry for enhancement of pump absorption; yellow part schematically represent possible input pump distribution. (b) Numerical simulation of pump power distribution between the two fibers in the GT-Wave two-fiber bundle for the case of straight arrangement and coiled and twisted fiber bundle. Used with permission from ref. [3] (©[2016] IEEE).

This tutorial lecture will review principles of efficient pump absorption in double-clad fibers, fiber fabrication and characterization, pump and signal combiners, large mode area fibers, as well as practical limitations, including transversal [6] and longitudinal [7] mode instabilities, and emerging technologies.

References

[1] M. N. Zervas and C. A. Codemard, "High power fiber lasers: A review," IEEE J. Sel. Topics Quantum Electron. **20**, 0904123 (2014).
[2] L. Dong and B. Samson, *Fiber Lasers: Basics, Technology and Applications*, (CRC Press, Taylor & Francis Group, Boca Raton, 2017).
[3] P. Koska, P. Peterka, and V. Doya, "Numerical modeling of pump absorption in coiled and twisted double-clad fibers," IEEE J. Sel. Topics in Quantum Electron. **22**, 4401508, (2016).
[4] P. Koska and P. Peterka, "Numerical analysis of pump propagation and absorption in specially tailored double-clad rare-earth doped fiber," Opt. Quantum Electron. **47**, 3181 (2015).
[5] P. Koska, P. Peterka, J. Aubrecht, O. Podrazky, F. Todorov, M. Becker, Y. Baravets, P. Honzatko, and I. Kasik, "Enhanced pump absorption efficiency in coiled and twisted double-clad thulium-doped fibers," Opt. Express **24**, 102 (2016).
[6] C. Jauregui, J. Limpert, and A. Tünnermann, "High-power fibre lasers," Nature Photonics **7**, 861 (2013).
[7] P. Peterka, P. Koška, and J. Čtyroký, "Reflectivity of superimposed Bragg gratings induced by longitudinal mode instabilities in fiber lasers," IEEE J. Sel. Topics Quantum Electron. **24**, 0902608 (2018).

Progress in Optical Parametric Oscillators

M. Ebrahim-Zadeh[1,2]

[1] *ICFO-Institut de Ciencies Fotoniques, Mediterranean Technology Park, 08860 Castelldefels, Barcelona, Spain*
[2] *Institucio Catalana de Recerca i Estudis Avancats (ICREA), Passeig Lluis Companys 23, Barcelona 08010, Spain*

Optical parametric oscillators are uniquely versatile sources of tunable coherent radiation in spectral regions inaccessible to conventional lasers [1]. The advent of novel nonlinear optical materials over the past two decades, combined with progress in laser pump sources, has had an unprecedented impact on OPOs, enabling advancement of this technology to new spectral and temporal domains. The rapid developments have led to the realization of a new generation of OPO sources capable of providing widely tunable radiation with unparalleled performance capabilities with regard to wavelength coverage, output power and efficiency, spatial, spectral and temporal coherence across vast spectral regions. Harnessing the superior linear and nonlinear optical properties of new birefringent and quasi-phase-matched materials, together with the application of innovative design concepts, combined with novel solid-state and fiber pump lasers, OPOs have been transformed from proof-of-concept laboratory prototypes to truly viable sources of tunable coherent radiation from ~250 nm in the UV up to ~12 μm in the deep-IR, and in all temporal domains from continuous-wave to few-cycle femtosecond time-scales [1-3]. The latest examples of progress in OPOs include broadband phase-locked pulse generation in picosecond time-scale [4], see Fig. 1, and widely tunable soliton generation in the femtosecond domain [5], shown in Fig. 2.

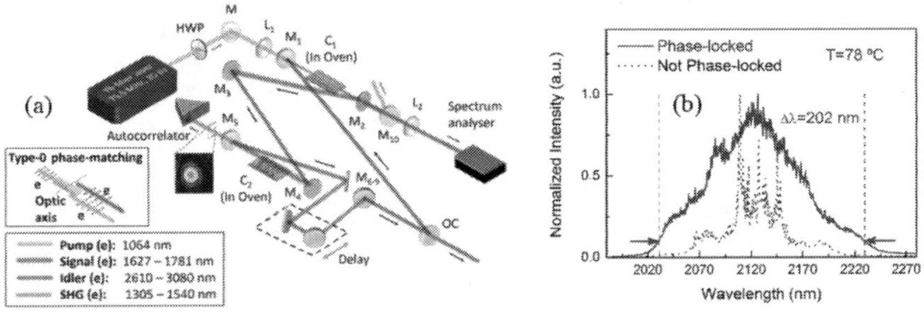

Fig. 1 (a) Yb-fiber-pumped phase-locked degenerate picosecond OPO based on MgO:PPLN for broadband generation in the near-infrared. (b) OPO output spectrum in the absence of phase-locking and in degenerate phase-locked state [4].

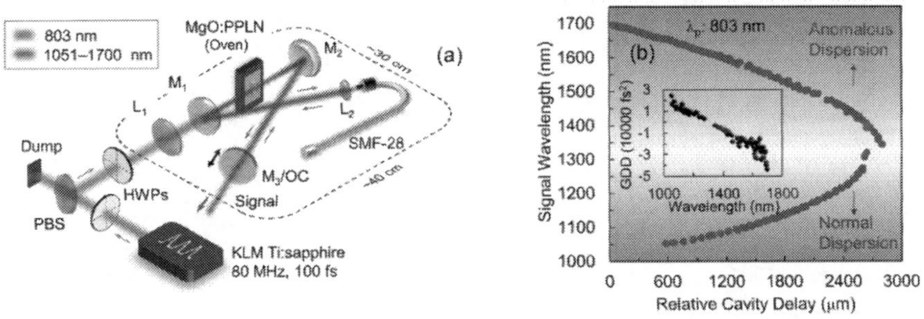

Fig. 2 (a) Ti:sapphire pumped soliton fibre-feedback OPO. (b) Cavity delay tuning of the OPO; inset: calculated net cavity GDD across the tuning range [5].

This lecture will provide an overview of parametric sources, from the basic design concepts to advanced systems, also highlighting the latest progress at the frontiers of OPO technology from the ultraviolet to the deep-infrared and from the continuous-wave to the ultrafast few-cycle operating time-scales. The lecture will include a description of nonlinear materials, innovative device architectures, applications of OPO sources in spectroscopy, imaging, environmental sensing, life sciences and biomedicine, and industrial development of OPO technology.

References

[1] M. Ebrahim-Zadeh and M. H. Dunn, "Optical parametric oscillators," in *Handbook of Optics IV*, 2nd ed., Optical Society of America, McGraw-Hill, New York, Ch. 22, pp. 22.01-22.72 (2001).
[2] M. Ebrahim-Zadeh, "Mid-infrared ultrafast and continuous-wave optical parametric oscillators," in *Solid-State Mid- Infrared Laser Sources*, I. T. Sorokina and K. L. Vodopyanov eds., Springer-Verlag, Berlin, Topics Appl. Phys. 89, 179-219 (2003).
[3] M. Ebrahim-Zadeh, "Continuous-wave optical parametric oscillators", in *Handbook of Optics IV*, 3rd ed., Optical Society of America, McGraw-Hill, New York, Ch. 17, pp. 17.1-17.33 (2010).
[4] B. Nandy, S. Chaitanya Kumar, M. Ebrahim-Zadeh, "Phase-locked picosecond optical parametric oscillator", Opt. Lett. **45**, 3981 (2020).
[5] C. F. O'Donnell, S. Chaitanya Kumar, T. Paoletta, M. Ebrahim-Zadeh, "Widely tunable femtosecond soliton generation in a fiber-feedback optical parametric oscillator", Optica 7, 426 (2020).

 https://doi.org/10.1051/epjconf/202024304001

Flexible all-PM NALM Yb:fiber laser design for low-noise frequency comb applications and single-cavity dual-comb generation

Oliver H. Heckl

Christian Doppler Laboratory for Mid-IR Spectroscopy and Semiconductor Optics, Faculty Center for Nano Structure Research, Faculty of Physics, University of Vienna, Boltzmanngasse 5, 1090 Vienna, Austria

Dual-comb systems traditionally consist of two mode-locked lasers actively stabilized to each other, leading to expensive and complex systems [1]. A very promising simplification is the single-cavity approach, where two pulse trains are generated within a single laser cavity [2]. Here, we present dual-comb generation from a single all-polarization-maintaining Ytterbium (Yb)-fiber laser via spectral subdivision. The laser is mode-locked using a nonlinear amplifying loop mirror (NALM) with a non-reciprocal phase bias. This versatile architecture allows for both single- and dual-comb operation in various intracavity group delay dispersion regimes.

In the single-comb operation mode, we have characterized the intensity noise, timing jitter and the free-running linewidth of the carrier-envelope-offset (CEO) frequency of five representative mode-locking states ranging from anomalous (-0.035 ps^2) to normal (+0.015 ps^2) intracavity dispersion [3]. We show that operation points far from the spontaneous emission peak of Yb (~1030 nm), but close to zero intracavity dispersion can be found, where the influence of pump noise is suppressed so strongly that the CEO-linewidth can be reduced to the single-digit kHz level at 1s integration time without any active stabilization.

Dual-color operation in such a laser is achieved by adding a thin blade-shaped beam block in the grating compressor where the light is spatially dispersed, see Fig. 1(a). The laser emits two pulse trains centered around 1030 nm and 1060 nm, respectively. By tuning the cavity dispersion, we can vary the difference in repetition rates from ~1 kHz to 10 kHz around a nominal repetition rate of 77 MHz, which corresponds to a maximum non-aliasing bandwidth of ~2.5 THz. The spectral overlap necessary for dual-comb operation is realized outside of the laser cavity by amplification and subsequent nonlinear broadening. The usability of this dual-comb setup is demonstrated by measuring the transmission of a Zinc Selenide (ZnSe) etalon.

The single-cavity dual-color/dual-comb all-PM Yb:fiber laser platform [4] presented here can easily be combined with power scaling stages and subsequent nonlinear wavelength conversion to expand its spectral operation range. Hence, such lasers show great potential as a versatile tool for many applications in spectroscopy and beyond.

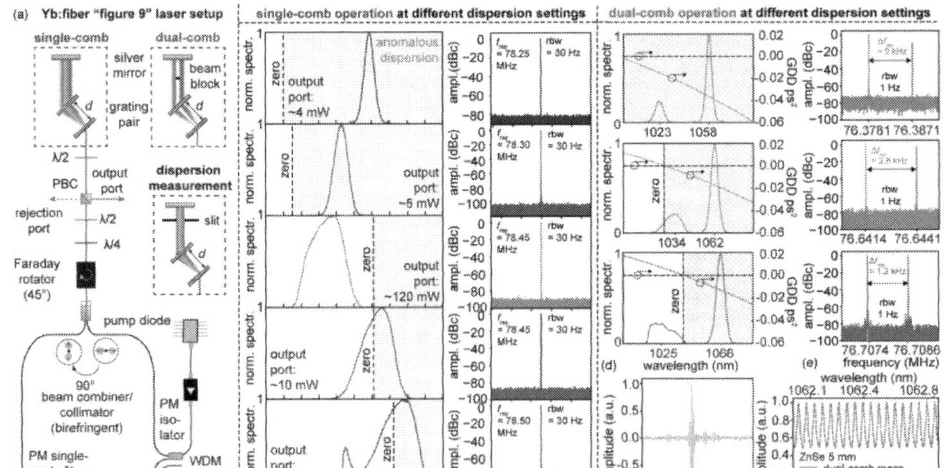

Figure 1: (a) Laser setup depicting the single- and dual-comb operation mode as well as the slit-method used for intra-cavity dispersion measurements. (b) Optical spectra of modelocking states obtained in single-comb operation at 5 different grating separations *d* and (c) corresponding measurement of the repetition rate frequency f_{rep}. (d) Optical spectra in dual-comb operation at 3 different grating separations and (e) corresponding measurement of the repetition rate frequencies $f_{rep,1}$ an $f_{rep,2}$. (f) Single-shot time domain interferogram of the configuration with Δf_{rep} =2.6 kHz corresponding to (g) transmission measurement of a 5-mm thick ZnSe (orange) and simulation (blue)

References:

[1] I. Coddington, N. Newbury, and W. Swann, "Dual-comb spectroscopy," *Optica*, vol. 3, no. 4, p. 414, 2016.

[2] T. Ideguchi, T. Nakamura, Y. Kobayashi, and K. Goda, "Kerr-lens mode-locked bidirectional dual-comb ring laser for broadband dual-comb spectroscopy," *Optica*, vol. 3, no. 7, p. 748, 2016.

[3] A. S. Mayer *et al.*, "Flexible all-PM NALM Yb:fiber laser design for frequency comb applications: operation regimes and their noise properties," *Opt. Express*, vol. 28, no. 13, pp. 18946–18968, 2020.

[4] J. Fellinger *et al.*, "Tunable dual-comb from an all-polarization-maintaining single-cavity dual-color Yb:fiber laser," *Opt. Express*, vol. 27, no. 20, p. 28062, Jun. 2019.

EPJ Web of Conferences **243**, 04002 (2020) https://doi.org/10.1051/epjconf/202024304002
EUROPHOTON 2020

A passively mode-locked Holmium fiber oscillator based on a Nonlinear Amplifying Loop Mirror operating at 2050 nm

Christoph Mahnke[1], Yuxuan Ma[1], Sarper Salman[1,2], Christoph M. Heyl[1,2] and Ingmar Hartl[1]
1. Deutsches Elektronen-Synchrotron DESY, Notkestraße 85, 22607 Hamburg, Germany
2. Helmholtz-Institute Jena, Fröbelstieg 3, 07743 Jena, Germany
christoph.mahnke@desy.de

In recent years, mode-locked fiber lasers have proven to be reliable sources for femto- and picosecond pulses for various applications like communications, metrology or biomedical applications [1]. We present first results on a Ho:fiber oscillator based on a Nonlinear Amplifying Loop Mirror (NALM) configuration, intended for seeding a Ho:YLF regenerative amplifier [2]. The NALM technique, when using polarization-maintaining (PM) fibers, allows self-starting, low noise, alignment-free, robust long term stable operation [1]. It is implemented here –to the best of our knowledge– for the first time using a Ho-fiber without Tm co-doping.

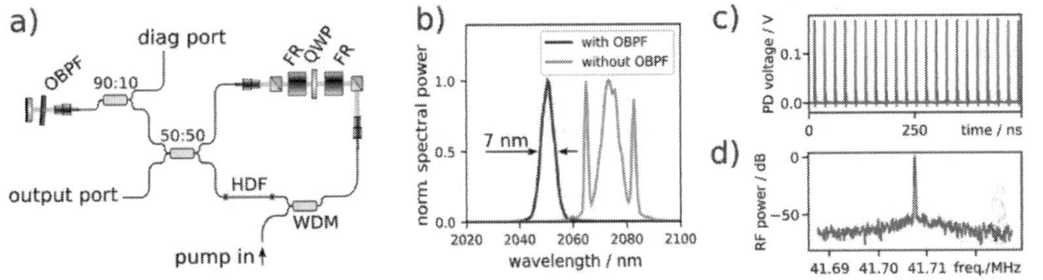

Fig.1: a) overview of the setup. b) optical spectra of mode-locked operation. c) pulse train in the time domain. d) radio-frequency spectrum (resolution bandwidth 100 Hz) around the repetition rate.

The schematic setup of the oscillator is shown in figure 1 a). It consists of a Nonlinear Amplifying Loop Mirror, a linear arm and a rejection port which are connected by a 50:50 fiber coupler. All passive fiber sections are PM fiber optimized for 2 um applications (Nufern PM1950).

The NALM loop is formed by a section of PM Holmium-doped gain fiber (Nufern PM-HDF-10/130-15FA) and additional sections of passive fiber. The asymmetric placement of the active fiber in the loop provides the difference in nonlinear phase shifts for clockwise and counter-clockwise propagation that is needed to introduce an intensity-dependent reflectivity of the NALM. An additional linear nonreciprocal phase shift is introduced in a short free space section with two Faraday rotators (FR) and a quarter wave plate (QWP), enabling self-starting mode-locking [3,4].

The linear arm of the oscillator contains a 10:90 coupler used for diagnostics and a free space-section with a small metallic end mirror mounted on a piezo actuator intended for cavity length stabilization. An optical bandpass filter (OBPF) can be placed in the free space section to select the operation wavelength. To pump the system, light from a commercial Thulium pump laser is coupled into the oscillator using a wavelength-division multiplexer (WDM).

Mode-locking of the oscillator is initiated when the power of the pump light is set to about 1.4 W. Typically, the laser starts pulsing in a multi-pulse modus. A stable, single pulse regime is reached when the pump power is reduced down to about 700 mW; mode-locking is lost at about 580 mW of pump power.

Without the OBPF in the linear arm, the oscillator shows a typical soliton spectrum with the characteristic Kelly sidebands (gray spectrum in Fig. 1b)). Placing and tilting the OBPF allows to adjust the wavelength to about 2050 nm (blue spectrum in Fig. 1b)), which is required for the intended seeding of a Ho:YLF amplifier. The OBPF greatly suppresses the Kelly sidebands. The bandwidth of the optical spectrum is about 7 nm, supporting pulse durations of about 1 ps. The repetition rate is set to 41.7 MHz and will be stabilized in the future using the piezo-actuated end mirror. Fig. 1c) shows the temporal trace of the pulses generated, recorded with a fast photo diode. The radio frequency spectrum in the vicinity of the repetition rate is depicted in Fig. 1d).

The output power from the output port is 4 mW, corresponding to a pulse energy of about 95 pJ. About 2 mW of power from the coupler in the linear arm is available for diagnostics.

References

[1] J. Kim and Y. Song, "Ultralow-noise mode-locked fiber lasers and frequency combs: principles, status, and applications," Adv. Opt. Photonics **8**, 465 (2016).
[2] P. Kroetz, et al. "Study on laser characteristics of Ho: YLF regenerative amplifiers: Operation regimes, gain dynamics, and highly stable operation points," Appl. Phys.B, **123**, 126 (2017).
[3] N. Kuse, et al. "All polarization-maintaining Er fiber-based optical frequency combs with nonlinear amplifying loop mirror," Opt. Express **24**, 3095 (2016).
[4] W. Hänsel et al., "All polarization-maintaining fiber laser architecture for robust femtosecond pulse generation," Appl. Phys. **B 123**, 41 (2017).

 https://doi.org/10.1051/epjconf/202024305001

Multi-Watt, mJ nanosecond pulses amplification in a Yb:LuLiF$_4$ single crystal fiber grown by micro-pulling-down

S. Pizzurro[1], F. Pirzio[1], S. Jun[3], A. Di Lieto[2], M. Tonelli[4], A. Agnesi[1]

1. Dipartimento di Ingegneria Industriale e dell'Informazione, Università di Pavia, via Ferrata 5, IT-27100, Pavia, Italy
2. NEST Istituto Nanoscienze-CNR and Dipartimento di Fisica, Università di Pisa, Largo B. Pontecorvo 3, IT-56127 Pisa, Italy
3. Gemmological Institute, China University of Geosciences, Wuhan 430074, China
4. Mega Materials s.r.l and Dipartimento di Fisica, Università di Pisa, Largo B. Pontecorvo 3, IT-56127 Pisa, Italy

Single crystal fibers (SCFs) grown by the micro-pulling-down (μ-PD) method represent a very attractive solution for power scaling of solid-state lasers owing to a long absorption length and a favorable surface-to-volume ratio that enhances thermal dissipation. μ-PD technique was successfully employed mainly in growing cubic crystals, most impressive results being obtained with Yb:YAG [1]. We recently demonstrated successful growth of a birefringent Yb-doped LuLiF$_4$ (LLF) SCF and cw-regime high-power amplification with excellent beam quality preservation even at very high (~ 90 W) absorbed pump power [2]. Birefringent SCFs are particularly attractive, as they provide a quite straightforward means of avoiding beam quality degradation at high thermal load due to depolarization. In this work we present what is, to the best of our knowledge, the first demonstration of pulse amplification in the nanosecond regime in a Yb-doped birefringent SCF amplifier.

The seeder for the experiment was a house-made, diode-pumped Yb:YLF passively Q-Switched laser emitting at 1020 nm and providing 110-ns, 180-μJ pulses at 5.2 kHz repetition rate (≈ 1-W average power), in an almost diffraction limited beam quality ($M^2 \approx 1.1$). The setup of the multi-pass amplifier is shown in Fig. 1(a). In this configuration, the seed beam polarization is rotated by the Quarter-Wave Plate (QWP) so that 1st and 4th pass exploit the amplifier gain for polarization parallel to Yb:LLF *c*-axis, whereas the 2nd and 3rd pass produce amplification along the *a*-axis. The seed beam was focused to $w_s \approx 140$ μm by the lens f$_2$, and spherical mirror R$_1$ re-imaged the seed in the amplifier.

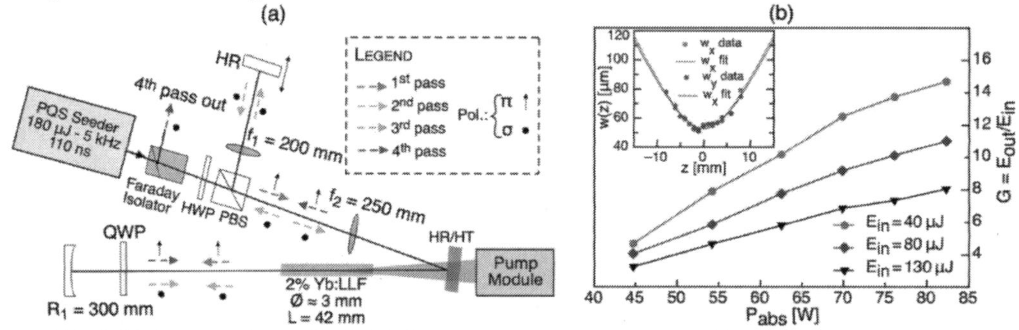

Fig. 1 (a) Experimental setup. HR: High-Reflectivity mirror at 1020 nm; HWP: Half-Wave Plate at 1020 nm; PBS: Polarizing Beam Splitter; HR/HT: flat dichroic highly reflective at 1020 nm and with High Transmittivity at 976 nm; QWP: Quarter-Wave Plate at 1020 nm. (b) Four passes amplification gain as a function of the absorbed pump power for different seeder energy. In the inset we show the result beam quality measurement carried out after amplification.

The Yb:LLF SCF was pumped by a 100-W maximum power, multimode fiber coupled (NA = 0.22, Ø = 105 μm) laser diode emitting at 976 nm. The pump beam was focused in the amplifier to a $w_p \approx 200$ μm, in order to match the pump beam confocal parameter to the 42-mm SCF length. The amplifier crystal was mounted in a water-cooled copper holder kept at 4 °C. A maximum pulse energy/average output power of 1.05 mJ/5.5 W was obtained at the maximum incident seeder energy/average power of 130 μJ/670 mW, corresponding to a four-passes gain G ≈ 8 (see Fig. 1(b)) and resulting in a maximum amplified pulse peak power of 10 kW. By fitting the experimental data, we estimate a saturation power of the amplifier of about 10 W and a four-passes small signal gain $G_0 \sim 40$. Seeder beam quality was very well preserved during amplification, even at the maximum absorbed pump power > 80 W. In the inset of Fig. 1(b) we show the beam quality measurement results after amplification and after we corrected the slight astigmatism of the amplified beam (already present in the seeder) with a telescope using cylindrical lenses, obtaining an almost perfectly circular beam with $M^2 = 1.1$.

References

[1] F. Lesparre, J. T. Gomes, X. Délen, I. Martial, J. Didierjean, W. Pallmann, B. Resan, M. Eckerle, T. Graf, M. A. Ahmed, F. Druon, F. Balembois, and P. Georges, "High-power Yb:YAG single-crystal fiber amplifiers for femtosecond lasers in cylindrical polarization," Opt. Lett. **40**, 2517–2520 (2015).
[2] F. Pirzio, S. Jun, S. Tacchini, A. Di Lieto, G. Piccinno, M. Tonelli, and A. Agnesi, "Multi-Watt amplification in a birefringent Yb:LuLiF$_4$ single crystal fiber grown by micro-pulling-down," Opt. Lett. **44**, 4095–4098 (2019).

Diode bar pumped, 0.5 mJ, sub-ns laser at 1.34 μm

***Kaloyan Georgiev*[1], Lyuben S. Petrov[1], *Desislava N. Georgieva*[1], *Anton Trifonov*[1, 2], *Ivan Buchvarov*[1]**
1. Department of Physics, Sofia University, 5 James Bourchier Blvd., Sofia, Bulgaria
2. IBPhotonics Ltd., Plovdivsko pole 19A,Sofia, Bulgaria

Compact lasers, emitting sub-ns, high energy and peak power pulses in wavelength near 1.3 μm spectral range are essential for wide variety of applications such as autonomous vehicle, remote sensing, frequency conversion and so on. Some of the recent and promising applications require both high pulse energy (~mJ) and peak power (~1 MW) at frequencies above 100 Hz. The diode-pumped short-cavity oscillators, working in passively Q-switching regime of operation, are now established as ultra-compact laser sources, capable to produce optical pulses with energy 1-1000 μJ and sub-ns duration with high beam quality [1, 2]. This is well developed mostly for Nd^{+3} doped active media emitting mainly at 1 μm transition $F_{3/2} \rightarrow {}^4I_{11/2}$ [1]. In addition, to achieve single transfer mode of operation at high excitation energy a fiber coupled diode with more than 100 W peak power was used for longitudinal pumping. However the transition ${}^4F_{3/2} \rightarrow {}^4I_{13/2}$ around 1.3 μm has substantially smaller emission cross which affects the formation of the leading edge of the Q-switched laser pulse and hence the obtaining of shorter pulse duration [4] and much higher pump energy is required. At present, diode laser bars with high fill factor can emit 200-300 W at pump wavelength in QCW operational mode, but the beam profile of a single diode laser bar is highly asymmetric to be used directly as a TEM_{00}-pump source.

In this work, we report a compact, laser diode bar pumped, single mode, passively Q-switched, Nd^{+3}:YAP based oscillator, providing TEM_{00} output at 1.34 μm, with 0.55 mJ, 730 ps pulses at 100 Hz repetition rate. A compact beam shaping scheme is developed for end pumping of 6 mm long Nd^{+3}:YAP crystal Fig. 1 (a).

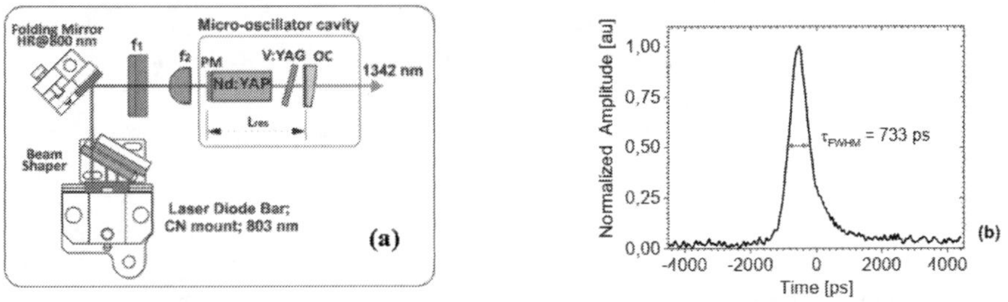

Fig. 1. Layout of the laser system **(a)** and oscilloscope trace of the Q-switched pulse **(b)**.

For pump source we use a TEC cooled, quasi-CW laser diode bar, emitting pulses on 803 nm (at 18°C) with maximal duty cycle of 4%. In order to obtain a compact design, the radiation from the diode bar was passed through a specially designed beam shaper, which partially symmetries the excitation laser beam in the x - y directions, perpendicular to the laser cavity axis. It is a modification of the idea proposed in [5,6]. The resulting beam is focused by two cylindrical lenses to a nearly circular pump area (~ 400 μm, $1/e^2$ width) in the active medium (ϕ4x6 mm, b-cut Nd^{+3}:YAP). The exact position of the lens with the shorter focal length f_2 along z-axis is adjusted by optimizing the minimum value of the laser threshold. The positioning of the lens f_1 has been done in a way that provides a single TM_{00} mode of operation of the laser oscillator.

The laser operation has been modeled numerically using three coupled differential equations for the passively Q-switched laser oscillator [3], in order to obtain laser pulses with maximum output energy and as shorter as possible pulse duration. Based on the theoretical analysis, we constructed a 12 mm long, plane-parallel laser resonator with 50% output coupler. For passive Q-switching, we use a V^{+3}:YAG saturable absorber with initial absorption T_0 = 52% for 1342 nm. Maximum output energy of 0.55 mJ at 1342 nm is obtained in Q-switching for 19 mJ absorbed pump energy. The pump pulse duration is 170 μs at 100 Hz repetition rate. At these resonator parameters, the measured Q-pulse duration is 733 ps (deconvoluted with system response function) with corresponding peak power of 0.75 MW (Fig 1 (b)). The output beam is near diffraction limited (M^2_x = 1.2) with 2.5 mrad divergence and linear polarization (polarization extinction ratio 1:200).

References

[1] J. J. Zayhowski , *"Passively Q-Switched Microchip Lasers"*, Chapter 1 in Solid - State Lasers and Applications, CRC Press (2006).
[2] Jan Sulc, D. D. S., et al. (2005). *"Q-switched Nd:YAG/V:YAG monolith microlaser"*. Proc.SPIE.
[3] J. J. Degnan, *"Optimization of passively Q-switched lasers"* in *IEEE Journal of Quantum Electronics*, vol. 31, no. 11, pp. 1890-1901, Nov. 1995.
[4] Sulc, J., et al.. *"Comparison of diode-side-pumped Nd:YAG and Nd:YAP laser"*, Proceedings of SPIE Vol. 5707 (2005).
[5] W. A. Clarkson and D. C. Hanna, "Two-mirror beam-shaping technique for high-power diode bars," Opt. Lett. 21, 375-377 (1996)
[6] Gunnar, RUSTAD, LIPPERT Espen and STENERSEN Knut. *"Beam Shaping of High Power Laser Diode Bars" FFI RAPPORT*, Conference Proceedings (2001).

SHG and SFG processes at a 100 kHz picosecond diode-pumped Yb:YAG thin-disk laser

H. Turcicova, O. Novak, J. Muzik, D. Stepankova, M. Smrz, A. Lucianetti, and T. Mocek
HiLASE Centre, Institute of Physics AS CR, Za Radnicí 828, 25241 Dolní Břežany, Czech Republic

Generation of harmonic frequencies is an efficient tool by which the spectral region of exploitable wavelengths of a laser can be substantially broadened. Besides the second harmonic, which is a prerequisite for further sum frequency processes, the UV and deep UV (below 300 nm, DUV) radiation is used for processing of different materials. The picosecond pulses enable short intense contact between the laser radiation and material. Such type of material processing shows specific features which cannot be attained otherwise (very precise microstructuring with sharp edges and high aspect ratio).

The fundamental beam is generated in a chirped pulse amplification laser system PERLA C running at 100 kHz [1]. It consists from a front end (fiber oscillator, fiber amplifiers) followed by a diode-pumped thin-disk Yb:YAG (1030 nm) regenerative amplifier. At present the 1-2 ps pulses are available up to 2 mJ energy. The SHG is realized in an LBO crystal, the wavelength of 515 nm is then doubled in a CLBO or BBO crystal to generate the DUV radiation of 257.5 nm. By SFG processes the third harmonic of 343 nm in an LBO crystal and the fifth harmonic of 206 nm in a CLBO or BBO crystal are generated. A challenging factor in the harmonics generation is the conversion efficiency related to the fundamental beam. Whereas the SHG from the fundamental beam can reach conversion around 70% [2], the SHG from the frequency doubled beam faces an adverse effect of the two-photon absorption leading to a conversion efficiency decrease, see e.g. [3]. The SFG process is used as $(2\omega + 1\omega)$ for the third harmonic generation and $(4\omega + 1\omega)$ for the fifth harmonics. The generation of DUV radiation meets challenging effects to be overcome: different group velocities of 1ω and 4ω pump beams inside the crystal and nonlinear absorption of both DUV photons. A mitigation to the group velocity mismatch is matching the group velocities by tilting the pulse fronts and noncollinear interaction of the beams inside the crystal [4]. The nonlinear absorption, however, is difficult to mitigate, the only way to lower its effect is to use low intensity beams together with higher crystal temperature, when the coefficient of the nonlinear absorption is getting lower, see e.g. [5]. The nonlinear absorption also presents a serious problem for DUV beams exiting the boxes containing protective atmosphere for crystals (nitrogen, argon). The exit windows should be produced from materials with low nonlinear absorption coefficient, such as MgF2, CaF2, LiF, or crystalline quartz [6].

In our contribution we present the up-to-date status of our harmonics generation system and its readiness for the users' experiments. At present our users have at a disposal 50W in 2nd, 25 W in 3rd, 10 W in 4th, and 1 W in 5th harmonics, in pulses of 1-2 ps duration.

References

[1] M. Smrz, M. Chyla, J. Muzik, S. S. Nagisetty, O. Novak, Compact, picosecond, kW-class thin disk laser PERLA for hi-tech industrial applications, MM Science, 2019, December, 3621.
[2] H. Fattahi, A. Alismail, H. Wang, et al., High-power, 1-ps, all-Yb:YAG thin-disk regenerative amplifier, Optics Letts. **41**(6), 1126 (2016).
[3] H. Turcicova, O. Novak, L. Roskot, et al., New observations on DUV radiation at 257 nm and 206 nm produced by a picosecond diode pumped thin-disk laser, Optics Express **27**(17), 24286, https://doi.org/10.1364/OE.27024286 (2019).
[4] B. Willenberg, F. Brunner, Ch. Phillips,and U. Keller, High-power picosecond deep-UV source via group velocity matched frequency conversion, Optica **7**(5), 485 (2020).
[5] M. Takahashi, A. Osada, A. Dergachev, et al., Effects of pulse rate and temperature on nonlinear absorption of pulsed 262-nm laser light in BBO, Jap. J. Appl. Phys. **49**, 080211 (2010).
[6] S. Patankar, S. T. Yang, J. D. Moody, et al., Two-photon absorption measurements of deep UV-transmissible materials at 213 nm, Applied Optics **56**(30), 8309.

 https://doi.org/10.1051/epjconf/202024306003

Infrared and self-frequency-doubling emission characteristics of diode-pumped Nd:LGSB laser crystal

Catalina-Alice Brandus[1], Madalin Greculeasa[1,2], Alin Broasca[1,2], Flavius Voicu[1], Lucian Gheorghe[1]

1. National Institute for Laser, Plasma and Radiation Physics, Laboratory of Solid-State Quantum Electronics, Magurele 077125, Romania
2. Doctoral School of Physics, University of Bucharest, Faculty of Physics, Magurele 077125, Ilfov, Romania

Visible laser sources are attractive for various applications, such as laser display, information storage, optical communications, or in medicine. In addition, a self-frequency-doubling (SFD) laser can be incorporated in a chip device, bringing advantages like stability, efficiency, low cost, compactness and simplicity. Different Yb^{3+}- and Nd^{3+}-doped bifunctional crystals have been developed [1], including Yb:YCOB [2], Nd:GdCOB [3], Nd:NLBO [4], or Nd:CTGS and Nd:STGS [5,6]. The need for efficient SFD laser devices justifies the high interest for discovering novel bifunctional materials with enhanced performances [7]. In this work, we report on infrared laser emission and first SFD results at 531 nm achieved with a 2.3-at.% Nd:LGSB bifunctional crystal, grown by the Czochralski method in our laboratory.

The laser medium was a 2.3-at.% Nd:LGSB negative uniaxial crystal, uncoated, with a thickness of 5.7 mm; it was cut for type I phase matching condition (θ= 35.3°, φ= 60°) at room temperature. The crystal was fixed in a copper mount whose temperature was carefully controlled, in order to ensure proper cooling during lasing, while keeping the phase matching condition for SFD. The optical resonator (Fig. 1) was a short (10-mm length) plane-concave configuration. The plane pump mirror (HRM) was coated high transmission (HT, T> 98%) at the 807-nm pump wavelength (λ_p) and high reflectivity (HR, R> 99.9%) at both fundamental (λ_ω= 1062 nm) and green ($\lambda_{2\omega}$= 531 nm) laser wavelengths. The concave out-coupling mirror (OCM) had a radius ρ= 100 mm. The pump (at λ_p= 807 nm) was done with a fiber-coupled diode laser, whose fiber end (100-µm core diameter, NA= 0.22) was imaged into Nd:LGSB with a 1:1 optical system made of two aspherical lenses.

The laser emission at λ_ω= 1062 nm was evaluated first, using OCM with different transmissions, T. The best performance was recorded for an OCM having T= 0.05. As shown in Fig. 2a), the Nd:LGSB crystal yielded 1.55 W output power (P_{out}) at 4.5-W absorbed pump power (P_{abs}); the slope efficiency, with respect to P_{abs}, was η_{sa}= 0.42. Saturation of P_{out} was observed for P_{abs} exceeding 3.9 W, indicating thermal effects in Nd:LGSB.

For the generation of green light by SFD, an OCM coated HR (R> 99.9% at 1062 nm) and HT (T> 95% at 531 nm) was used. The green output power reached P_{out}= 13.2 mW for P_{abs}= 4.31 W at 807 nm [Fig. 2b)]. The spectrum was centred at $\lambda_{2\omega}$= 531 nm, having a bandwidth (FWHM definition) of 1.3 nm.

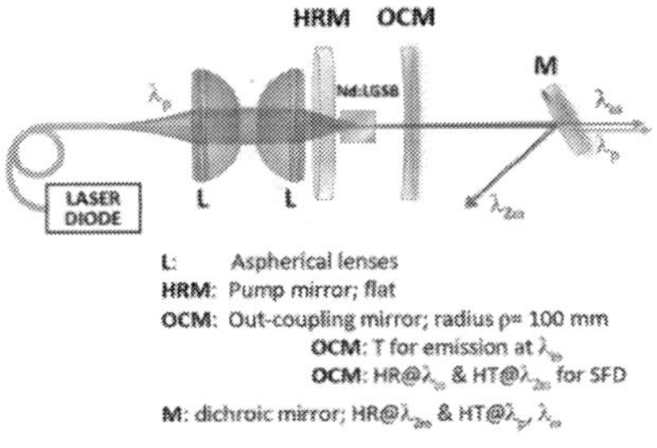

Fig. 1 The experimental set-up, general view.

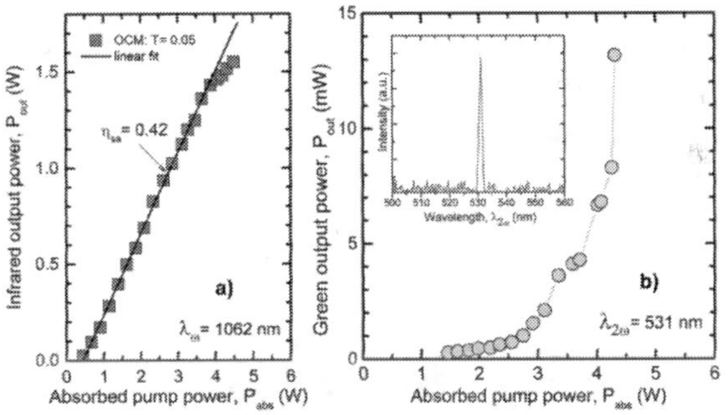

Fig. 2 Output power vs. aborbed pump power for emission at **a)** λ_ω= 1062 nm and **b)** SFD at $\lambda_{2\omega}$= 531 nm. Inset shows the emission spectrum at 531 nm.

In conclusion, we report on first SFD experiments performed on Nd:LGSB bifunctional laser crystal, from which 13.2 mW output power of green light at 531 nm was obtained. The results compare well with previous reports on other Nd-based bifunctional media [4-6]. Improved SFD performances are expected by optimizing the Nd:LGSB medium characteristics (doping level and length), by coating the crystal, as well as by designing a new laser resonator.

Financing: Projects 119/2017, PN-III-P4-ID-PCE-2016-0853 and NUCLEU 16N/2019, Ministry of Research and Innovation, Romania.

References

[1] A. Brenier, D. Jaque, A. Majchrowski, Opt. Mater. **28**(4), 310-323 (2006).
[2] F. Khaled, P. Loiseau, G. Aka, and L. Gheorghe, Opt. Lett. **41**(15), 3607-3610 (2016).
[3] H. Yu, N. Zong, Z. Pan, H. Zhang, J. Wang, Z. Wang, and Z. Xu, Opt. Lett. **36**(19), 3852-3854 (2011).
[4] J. Zhang, H. Yu, Y. Li, L. Han, Y. Wu and H. Zhang, Opt. Lett. **37**(17), 3501-3503 (2012).
[5] F. Chen, Y. Zhou, F. Yu, Y. Liu, S. Hou, H. Wang, Z. Wang, X. Duan, S. Guo, X. Zhao, J. Alloy. Compd. **651**, 475-478 (2015).
[6] F. Chen, L. Wang, X. Wang, X. Cheng, F. Yu, Z. Wang, X. Zhao, Opt. Mater. **73**, 33-37 (2017).
[7] L. Gheorghe, A. Broasca, M. Greculeasa, F. Voicu, G. Stanciu, S. Hau, G. Croitoru, C. A. Brandus, C. Gheorghe, F. Khaled, P. Loiseau, G. Aka, Opt. Mater. X **7**, 100052 (2020).

New Yb:LYSB bifunctional crystal for efficient near-infrared laser emission and self-frequency doubling conversion

Alin Broasca[1,2], Madalin Greculeasa[1,2], Flavius Voicu[1], Stefania Hau[1], Gabriela Croitoru[1], Cristina Gheorghe[1], Nicolaie Pavel[1], Lucian Gheorghe[1]

1. National Institute for Laser, Plasma and Radiation Physics, Laboratory of Solid-State Quantum Electronics, Magurele 077125, Romania
2. Doctoral School of Physics, University of Bucharest, Faculty of Physics, Magurele 077125, Romania

Nowadays, solid-state lasers with emission in the visible (VIS) spectral range are used in a large variety of applications, from manufacturing and materials processing technologies to medicine, optical communication or military applications. A widely used method for the obtaining of efficient laser sources in VIS range is based on frequency conversion of solid-state near-infrared laser emission by nonlinear optical (NLO) processes, such as second harmonic generation (SHG) or self-frequency doubling (SFD).

In this work, we report on the development of Yb:LYSB as new bifunctional laser and NLO crystal. Based on our experience on growing undoped $La_xY_ySc_{4-x-y}(BO_3)_4$ - LYSB - type crystals [1], a special heating assembly that allows an effective control of both B_2O_3 evaporation and thermal gradients over the crucible and in the melt, was used for the growth of the incongruent melting Yb:LYSB crystal by the Czochralski method (Fig. 1). The XRD analysis revealed the existence of the single trigonal phase, space group *R32*; the unit cell parameters were determined to be a= 9.8114(4) Å and c= 7.9808(9) Å. The chemical composition of the crystal was found to be $La_{0.78}Y_{0.32}Yb_{0.04}Sc_{2.86}(BO_3)_4$. The segregation coefficient of Yb^{3+} ions in the LYSB crystal matrix was determined to be $k_{eff} = 0.8$, corresponding to an Yb^{3+}-doping concentration of 4 at.% in the grown crystal. The absorption cross-section, σ_a at 980 nm for σ-polarization is about 0.6×10^{-20} cm^2 with a bandwidth (FWHM definition) of 33 nm, being ~1.65 times larger than that of Yb:YAB crystal [2]. In the case of π-polarization, the highest σ_a was determined to be 0.37×10^{-20} cm^2 at 904 nm, being close to the value reported for Yb:LGSB crystal [3]. Furthermore, spectroscopic studies revealed an intrinsic disorder due to the presence of Yb^{3+} ions in the trigonal prismatic sites occupied by La^{3+} and Y^{3+} ions and also in the octahedral sites occupied by Sc^{3+} ions.

Efficient laser emission at the fundamental wavelength λ_{em}= 1028 nm was obtained with a *c*-cut 4-at.% Yb:LYSB uncoated medium, having a thickness of 3.5 mm. The pump was done at 971.5 nm with a fiber-coupled (100-μm diameter, NA= 0.22) diode laser. A short (8-mm length) linear plane-plane resonator was used in the experiments. With an out-coupling mirror having transmission T= 0.01, the Yb:LYSB medium yielded laser pulses with maximum energy E_p= 1.5 mJ at 4.25 mJ energy of the absorbed pump pulse, E_{abs}; thus, overall optical-to-optical efficiency, with respect to E_{abs}, was η_{oa}= 0.35. The slope efficiency was as high as η_{sa}= 0.57 (Fig. 2), proving the good optical quality of Yb:LYSB grown crystal.

Fig. 1 Yb:LYSB as-grown crystal.

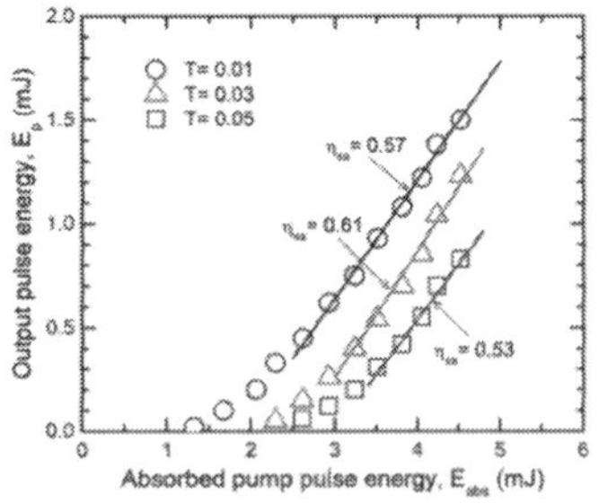

Fig. 2 Laser pulse energy, E_p vs. absorbed energy of the pump pulse, E_{abs}. T is the out-coupling mirror transmission.

The NLO properties of Yb:LYSB crystal were determined from the refractive indices measurements. The effective nonlinear coefficient, d_{eff} for type I SHG of λ_{em}= 1028 nm was determined to be 1.08 pm/V, confirming the good NLO properties of huntite-type Yb:LYSB crystal. Further experiments to determine the full potential of Yb:LYSB crystal as bifunctional laser and NLO crystal in SFD configuration are now in progress.

Financing: Projects 119/2017, PN-III-P4-ID-PCE-2016-0853 and NUCLEU 16N/2019, Ministry of Research and Innovation, Romania.

References

[1] L. Gheorghe, M. Greculeasa, A. Broasca, F. Voicu, G. Stanciu, K. N. Belikov, E. Yu. Bryleva, O. Gaiduk, "Incongruent Melting $La_xY_ySc_{4-x-y}(BO_3)_4$: LYSB Nonlinear Optical Crystal Grown by the Czochralski Method," ACS Appl. Mater. Interfaces **11**(23), 20987-20994 (2019).
[2] P. Wang, J.M. Dawes, P. Dekker, D.S. Knowles, J.A. Piper, B. Lu, "Growth and evaluation of ytterbium-doped yttrium aluminum borate as a potential self-doubling laser crystal," J. Opt. Soc. Am. B **16**(1), 63-69 (1999).
[3] F. Khaled, P. Loiseau, F. Voicu, A. Achim, S. Hau, C. Gheorghe, G. Croitoru, N. Pavel, L. Gheorghe, G. Aka, "Spectroscopic properties and laser performances of Yb:LGSB nonlinear optical crystal," J. Alloy. Compd. **688**(Part A), 510-517 (2016).

Enhancement of the laser emission efficiency of Yb:Y_2O_3 ceramics via multi-step sintering method fabrication

George Stanciu, Flavius Voicu, Catalina-Alice Brandus, Cristina Tihon, Stefania Hau, Cristina Gheorghe, Gabriela Croitoru, Lucian Gheorghe

National Institute for Laser, Plasma and Radiation Physics, Laboratory of Solid-State Quantum Electronics, Magurele 077125, Romania

Y_2O_3 sesquioxide is a promising host material for solid-state lasers due to its high thermal conductivity (~13.5 $Wm^{-1}K^{-1}$), wide transparency range (0.2 ÷ 8 μm), high refractive index (1.9 to 2.1 in the 300 ÷ 800 nm spectral range), or its low cut-off phonon energy (380 cm^{-1}) [1]. Furthermore, Yb^{3+}-doped Y_2O_3 ceramics have attracted great attention due to their relatively long fluorescence lifetime, high quantum efficiency and broad absorption and emission bands [1,2]. However, the growth of large Y_2O_3 crystals from the melt is a difficult task due to the high melting point (>2400°C). By using ceramic processing technology, Y_2O_3 ceramics can be fabricated at about two-thirds of the melting temperature, which greatly reduces the cost and fabrication period.

In this work, 5.0 at.% Yb:Y_2O_3 transparent ceramics were obtained by solid-state reaction and multi-step sintering (MSS) method using an optimal selected content of 9.0 at.% La_2O_3 and 2.0 at.% ZrO_2 as sintering additives. Based on the chemical formula $(Yb_{0.05}La_{0.09}Zr_{0.02}Y_{0.84})_2O_3$, the raw powders were blended and ball milled in ethanol with agate balls and then dried and sieved. The powder mixture was pressed into green bodies using a stainless steel die set, and then isostatically pressed at 250 MPa to enhance the green body's strength. The two-step intermediate sintering was conducted in air, as follows [3]. First, the temperature was increased from room temperature to a relatively high-temperature of 1450°C, with a heating rate of 10°C/min and maintained at this temperature for 1 h. After that, the temperature was rapidly decreased to 1000°C where it was held for 40 h. Finally, the samples were sintered at 1760°C for 12 h under vacuum (10^{-4} Pa) in a furnace with a tungsten-molybdenum chamber. For comparison, the compacted green bodies were also densified by the normal vacuum sintering method, at 1760°C. The as-sintered ceramics were further annealed in air at 1300°C, for 20 h, to remove oxygen vacancies. Finally, the ceramic samples were mirror-polished to a thickness of 1.5 mm.

The phase identification revealed that all sintered ceramics have a cubic structure similar to pure Y_2O_3. The microstructure presents a good homogeneity for all ceramic samples, with the average grain size of about 15 μm. Furthermore, it was observed that the grains become more uniform in the case of Yb:Y_2O_3 ceramic obtained by the MSS method. As a result, a high transmission of 78.5% at the wavelength of 1100 nm was recorded for the Yb:Y_2O_3 ceramic fabricated by the MSS method (Fig. 1). Laser emission at 1.03 μm was obtained in all 5.0 at.% Yb:Y_2O_3 uncoated ceramic media under quasi-continuous wave pumping at 971 nm with a fiber-coupled (100-μm diameter and NA = 0.22) diode laser. The optical resonator was a short (10-mm length) plane-plane mirror configuration. As shown in Fig. 2, laser pulses with 1.12 mJ energy for absorbed pump pulse energy E_{abs} = 6.35 mJ at 971 nm, were obtained from the Yb:Y_2O_3 ceramic densified by the MSS method; the slope efficiency, with respect to E_{abs}, reached 33%. The Yb:Y_2O_3 ceramic fabricated by the normal sintering method yielded laser emission with a lower slope efficiency of 29%.

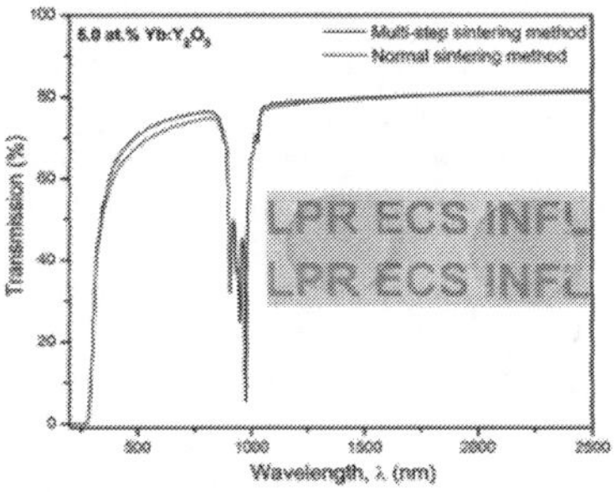

Fig. 1 Transmission spectra of the mirror-polished 5.0 at.% Yb:Y_2O_3 transparent ceramics.

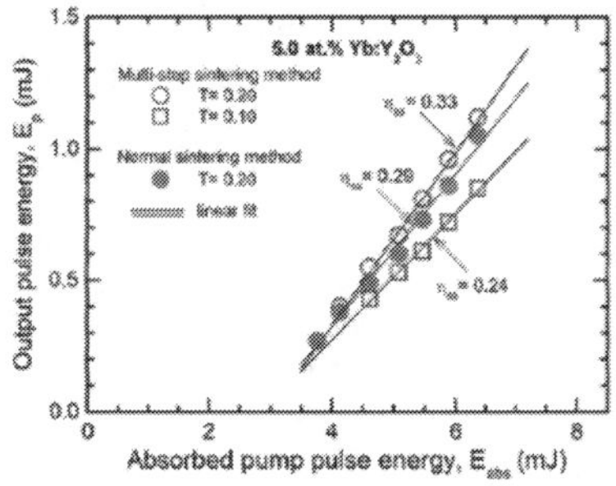

Fig. 2 Laser pulse energy, E_p versus absorbed energy of the pump pulse energy, E_{abs}. T: out-coupling mirror transmission.

Next investigations will focus on obtaining Yb:Y_2O_3 ceramics with better laser emission efficiency, as well as on the evaluation of their laser emission performances in different operating regimes.

Financing: Project NUCLEU 16N/2019, Romanian Ministry of Research and Innovation and project 654148, Laserlab-Europe EU-H2020.

References

[1] X. Hou, S. Zhou, W. Li, Y. Li, H. Lin, H. Teng, T. Jia, "Investigation of the spectroscopic properties of highly transparent Yb:$(Y_{0.97}Zr_{0.03})_2O_3$ ceramic," Opt. Mater. **32**, 1435-1440 (2010).

[2] J. Kong, J. Lu, K. Takaichi, T. Uematsu, K. Ueda, D. Y. Tang, D. Y. Shen, H. Yagi, T. Yanagitani, and A. A. Kaminskii, "Diode-pumped Yb:Y_2O_3 ceramic laser," App. Phy. Lett. **82**(16), 2556-2558 (2003).

[3] G. Stanciu, L. Gheorghe, F. Voicu, S. Hau, C. Gheorghe, G. Croitoru, M. Enculescu, R. P. Yavetskiy, "Highly transparent Yb:Y_2O_3 ceramics obtained by solid-state reaction and combined sintering procedures," Ceram. Int. **45**, 3217-3222 (2019).

Ultra-low-noise ultrafast fiber lasers

Alexander M. Heidt
Institute of Applied Physics, University of Bern, Sidlerstrasse 5, 3012 Bern, Switzerland

Femtosecond lasers based on mature Erbium-doped fiber (Er:fiber) technology have become the workhorse for ultrafast photonics research and applications due their stability, compactness, and ease of use. Commercial ultrafast Er:fiber lasers are available that have been developed specifically to exhibit ultra-low amplitude and phase noise, exhibiting relative intensity noise (RIN) values far below 0.1%, making them essential tools for demanding applications such as ultrafast spectroscopy or precision metrology [1].

A convenient approach of extending the excellent performance of Er:fiber lasers to other wavebands, e.g. into the 2 μm regime, is their nonlinear spectral broadening in nonlinear fibers, followed by amplification in a Thulium- or Holmium-doped fiber amplifier system. However, previous implementations using conventional nonlinear fibers pumped in the anomalous dispersion (AD) regime suffered from the generation of excess noise via quantum, technical, and polarization noise amplification by the involved nonlinear processes, resulting in RIN values in the range 0.3-0.7%, i.e. about an order of magnitude higher than the corresponding Er:fiber system This excess noise has recently been identified as the major performance limiting factor for the further development of high-power frequency comb sources at 2 μm [2].

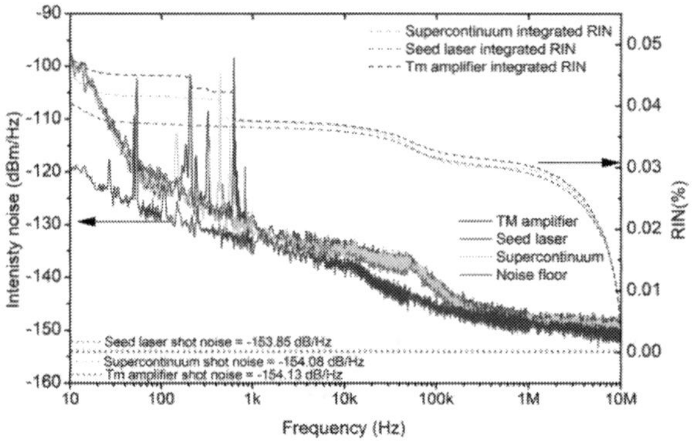

Fig. 1: Measured noise characteristics of the Er:fiber seed laser (red), ANDi supercontinuum (green), and Tm:fiber amplifier (blue) [5].

In this contribution I discuss the conditions required for suppressing this excess noise generation during nonlinear spectral broadening, present specialty optical fibers specifically designed for this purpose, and review the latest developments in their application aimed at realizing the next generation of ultra-low noise frequency combs and ultrashort pulse sources in the 2 μm spectral region. It is shown that ultra-low noise nonlinear spectral broadening and supercontinuum (SC) generation is enabled by all-normal dispersion (ANDi) fibers, which effectively suppress quantum and polarization noise amplification and lead to nonlinear dynamics that are extremely robust against technical noise [3,4]. The application of these ANDi SC sources for the coherent seeding of ultrafast Thulium- / Holmium-doped fiber amplifiers is demonstrated to result in an order-of-magnitude reduction of RIN to < 0.05%, i.e. virtually identical to the driving Er:fiber laser (Fig. 1) [5]. Therefore, neither the nonlinear spectral broadening nor the amplification process introduces significant excess noise, highlighting the potential of ANDi SC seed sources for the next generation of ultra-low noise, broadband, high-power ultrafast fiber amplifiers and frequency combs.

References

[1] D. Brida, G. Krauss, A. Sell, and A. Leitenstorfer, "Ultrabroadband Er:fiber lasers," Laser Photonics Rev. **8**, 409 (2014).
[2] C. Gaida *et al.*, "High-power frequency comb at 2 μm wavelength emitted by a Tm-doped fiber laser system," Opt. Lett. **43**, 5178 (2018).
[3] A. M. Heidt, J. S. Feehan, J. H. V. Price, and T. Feurer, "Limits of coherent supercontinuum generation in normal dispersion fibers," J. Opt. Soc. Am. B **34**, 764–775 (2017).
[4] B. Sierro and A. M. Heidt, "Noise amplification in all-normal dispersion fiber supercontinuum generation and its impact in ultrafast photonics applications," OSA Continuum (in press).
[5] A. Rampur, Y. Stepanenko, G. Stepniewski, T. Kardaś, D. Dobrakowski, D.-M. Spangenberg, T. Feurer, A. Heidt, and M. Klimczak, "Ultra low-noise coherent supercontinuum amplification and compression below 100 fs in an all-fiber polarization-maintaining thulium fiber amplifier," Opt. Express **27**, 35041–35051 (2019).

EPJ Web of Conferences **243**, 07002 (2020) https://doi.org/10.1051/epjconf/202024307002
EUROPHOTON 2020

Research on amplification of ultrashort laser pulses at 1.03 μm in gain-managed nonlinearity regime

Dorota Tomaszewska[1], **Grzegorz Soboń**[1]

1. Laser & Fiber Electronics Group, Wroclaw University of Science and Technology, Wybrzeze Wyspianskiego 27, 50-370 Wroclaw, Poland

Ultrafast laser sources with and high pulse energy are a topic of interest both in science and industry, especially when their wavelength are not reachable in traditional way. Wavelengths that differs from conventional gain bandwidths for fiber lasers are a great basis for nonlinear effects such as different frequency generation or pumps for mid-infrared parametric oscillators. Recently the group of F. Wise reported a new regime of amplification called gain-managed nonlinearity (GMN) [1]. In this regime both pulse and gain spectra evolve and reshape one another, which allows to generate a bandwidth that exceeds the gain spectrum for Ytterbium-doped fibers. It is shown that the along with the amplification process of the self-similar pulse, it experiences nonlinear deformation and pulse shifts towards longer wavelengths [1]. This point is considered to be the beginning of the GMN regime. The pulse then continues to broaden in spectral and time domain and its energy increases. Here, we demonstrate a simple fiber amplifier setup which delivers pulses with energy up to 71 nJ, which corresponds to 2.17 W of average power, with a central wavelength of 1083 nm and 45 nm of width.

The experimental setup of the source is depicted in Fig. 1(a). As a seed laser we used fiber laser with a repetition rate of 30.44 MHz, working at the central wavelength of 1031 nm and 11 nm of width (Fig. 1(b)). The compressed pulse duration was 312 fs at average power of 68.3 mW. Pulses from the seed were directed into the amplifier, which consisted of 9-meter-long double-clad Ytterbium-doped polarization maintaining (PM) fiber with a 10-μm core diameter. It was pumped by a laser diode working at 918 nm. The spectra obtained after GMN amplification are depicted in Fig. 1(c). While increasing the pump power, the spectrum is broadening and spanning beyond the gain bandwidth of Yb-doped fibers. The maximal average output power of the amplifier of 3.25 W was achieved at 6.94 W of pump power, which corresponds to pulse energy of 107 nJ. The central wavelength of 1115 nm and width of 92 nm was achieved for this pulse. For the pulse energy of 46 nJ or higher, Raman scattering is observed. The last spectrum not affected by Raman scattering had the average power of 0.995 W (32 nJ) for 3.32 W of pump power. The central wavelength for this pulse was 1083 nm and the width was 45 nm. To the best of our knowledge, these are the broadest spectra achieved so far from a single-stage gain-managed amplifier.

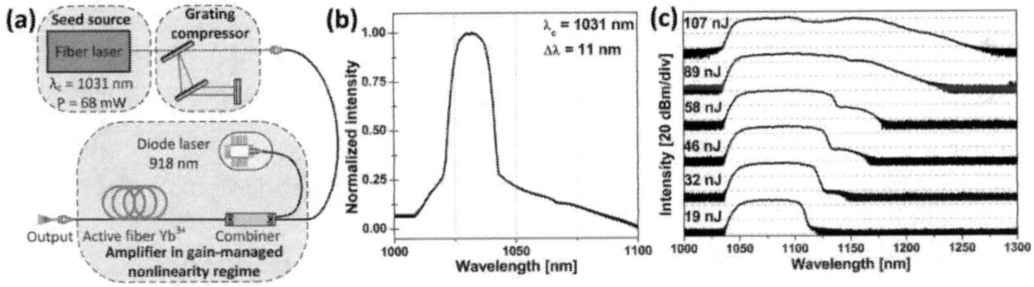

Fig. 1. Experimental setup of the GMN amplifier (a). Optical spectrum of the seed laser (b). Optical spectra of seed laser amplified in GMN regime (c).

Summarizing, we demonstrated an amplification process in a new gain-managed regime, which can deliver pulses with high energy and spectrum broader than a gain bandwidth. In the next step the presented source will be compressed with a grating compressor and used as pump for nonlinear processes [2], e.g. mid-infrared parametric generation. The work was funded by the Polish Ministry of Science and Higher Education (DG2018 006648).

References

[1] P. Sidorenko et al., "Nonlinear ultrafast fiber amplifiers beyond the gain-narrowing limit", Optica **6**, 1328-1333 (2019).
[2] A. Schliesser et al., "Mid-infrared frequency combs", Nat. Photonics **6**, 440-449 (2012).

Lasers and Optics for the Laser Interferometer Space Antenna (LISA)

Nelson Christensen, Artemis, OCA, Nice, France

The Laser Interferometer Space Antenna (LISA) will be a large-scale space mission designed to detect one of the most elusive phenomena in physics and astronomy, gravitational waves. With LISA we will be able to observe the entire universe directly with gravitational waves, learning about the formation of structure and galaxies, stellar evolution, the early universe, and the structure and nature of spacetime itself.

The era of gravitational wave astronomy has begun with the direct detections of gravitational waves from merging binary systems containing black holes and neutron stars by LIGO and Virgo. On Earth, observations are limited to objects with masses a few 10's of solar masses, which produce detectable signals in the 10 Hz to 1 kHz band. Larger mass sources, for example from the mergers of massive black holes at the centers of galaxies, will produce signals at much lower frequencies. These low frequency signals will be undetectable on Earth due to numerous noise limitations, seismic noise being one.

The European Space Agency (ESA) leads the mission, and consequently numerous European countries are contributing. NASA (USA) is a major collaborator, as well contributions from JAXA (Japan). The LISA Consortium consists of scientists from around the world. LISA will be the first space-based gravitational wave observatory. LISA will consist of three identical spacecraft in a triangular formation separated by 2.5 million km, following Earth in its solar orbit. Launch is planned for 2034.

LISA will be observing the gravitational wave universe in the frequency band from below 10^{-4} Hz to above 10^{-1} Hz. The observations will be made through the use of six active laser links between three spacecraft. These will be continuously operating heterodyne laser interferometers with pm/Hz$^{1/2}$ sensitivity. The light for each link will be from well-stabilized lasers at 1064 nm delivering a power of 2 W. However, only ~ 100 pW of power will be detected after the 2.5 Mkm propagation.

The relative arm-length measurements will be conducted via laser interferometry between freely floating test masses inside drag-free spacecraft. The test masses and the associated gravitational reference sensors are essentially identical to those recently flown on the successful LISA Pathfinder Mission. The test masses will follow their geodesic trajectories with sub femto-g/Hz$^{1/2}$ spurious acceleration. Two test masses will be in free-fall inside each spacecraft, with each one serving as a geodesic reference end mirror for a single arm of an interferometer. Each spacecraft must follow the two test masses along each of the two defined interferometry axes; this is based on local interferometric position measurements. The test masses will be electrostatically suspended to the spacecraft along the other degrees of freedom; the control will be implemented by a combination of interferometric and capacitive position readouts.

This talk will summarize the LISA Mission, with a special emphasis on the laser, optical and interferometeric systems. Sources of noise that will limit the measurements will be discussed. Also described will be the current state of the LISA Mission, and the activity plans up to the expected 2034 launch.

EPJ Web of Conferences **243**, 08002 (2020) https://doi.org/10.1051/epjconf/202024308002
EUROPHOTON 2020

High Resolution Mid-Infrared Up-Conversion LIDAR

Max Widarsson[1], Markus Henriksson[2]. Patrick Mutter[1], Carlota Canalias[1], Valdas Pasiskevicius[1], and Fredrik Laurell[1]
[1]*Department of Applied Physics, Royal Institute of Technology, Roslagstullsbacken 21, 106 91 Stockholm, Sweden*
[2]*Swedish Defence Research Agency, Olaus Magnus Väg 42, 583 30 Linköping, Sweden*

Light in the mid-infrared (MIR) has a higher transmission through certain media, such as fog and smoke [1], than visible/near-infrared due to the lower scattering probability. Therefore it is of interest to have LIDARs operating in this spectral region, specifically utilizing photon counting, which yields high sensitivity. However, most photon counting LIDAR systems operate in the near infrared [2-4]. This is due to a lack of good detectors in the MIR.

By up-converting the MIR photons to the working regime of established Si-detectors, it is possible to reap the benefits from both the MIR and good detectors. This technique has previously successfully demonstrated detection of MIR both through a single pass configuration [5] or when the nonlinear crystal was inside a laser cavity [6].

This experiment was performed to extend the wavelength of the up-conversion MIR LIDAR to 2.4 μm and achieve high spatial resolution compared to previous up-conversion photon counting LIDARs [7,8]. In this work the nonlinear up-conversion was done inside a $Nd:YVO_4$ cavity lasing at 1064 nm generating light at 737 nm.

The LIDAR was performed with the time of flight technique. To do this there were three main components of the setup, the MIR pulses, the conversion cavity and the detector for the up-converted light. The MIR pulses were reflected off a microscope slide, up-converted in the conversion cavity, spectrally filtered and detected. The MIR pulses had a temporal width below 100 fs with a spectral bandwidth of roughly 300 nm centered around 2.4 μm with a repetition rate of 1 kHz and pulse energy in the pJ regime.

A 10 mm long (9 mm grating length) periodically-poled rubidium-doped $KTiOPO_4$ (PPRKTP) with a 23.5 μm period was fabricated in-house and used for the up-conversion. The crystal was inserted into a straight $Nd:YVO_4$ laser cavity pumped by a 2.5 W 808 nm laser diode. The PPRKTP crystal was uncoated which resulted in a relatively low intra-cavity power of 5 W at 1064 nm.

A conventional silicon single-photon avalanche photodiode (Si-SPAD) with a detection efficiency of 23 % at 737 nm was used in combination with a time tagging unit with bin widths down to 4 ps.

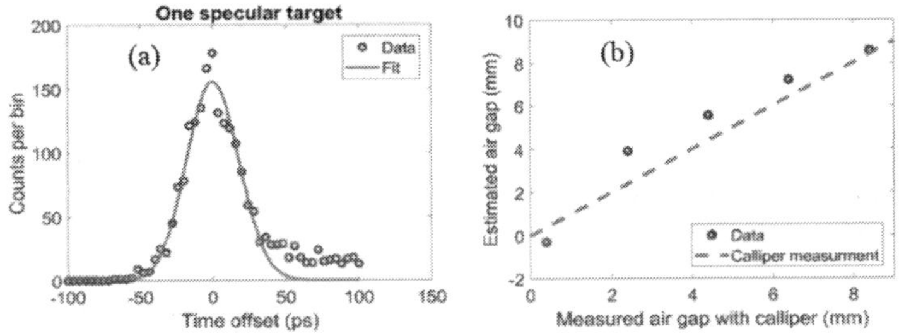

Fig. 1: (a) The measured signal from one specular target. The FWHM of the trace is 42 ps. The measured data is shown as circles and the solid line is a Gaussian fit. (b) Estimated air gap between the two targets.

To test the resolution of the system LIDAR traces were collected for multiple separations between two microscope slides varying from 0.4 mm to 8.4 mm of air and 1 mm of glass (thickness of the microscope slide).

The trace from a single reflection, as well as a Gaussian fit, can be seen in Fig 2 (a). The full width half maximum (FWHM) of a single reflection was only 42 ps. In Fig 2 (b) the estimated, through LIDAR, and the measured, by a caliper, air gap between the two microscope slides can be seen. An air gap separation of merely 0.4 mm (1.9 mm optical thickness) yielded a FWHM of 55 ps, which is 30 % broader than for a single target.

References

1. R. W. Bukowski, "Smoke measurements in large- and small-scale fire testing - Part I," Fire Technol. **15**(3), 173–179 (1979).
2. M. Henriksson, L. Allard, P- Jonsson, "Panoramic single-photon counting 3D lidar" Proc. SPIE **10796**, 1079606 (2018).
3. J. Qiu, H. Xia, M. Shangguan, X. Dou, M. Li, Chong Wang, Xiang Shang, Shengfu Lin, and Jianjiang Liu, "Micro-pulse polarization lidar at 1.5 μm using a single superconducting nanowire single-photon detector," Opt. Lett. **42**, 4454-4457 (2017).
4. R. E. Warburton, A. McCarthy, A. M. Wallace, S. Hernandez-Marin, R. H. Hadfield, S. Woo Nam, and G. S. Buller, "Subcentimeter depth resolution using a single-photon counting time-of-flight laser ranging system at 1550 nm wavelength," Opt. Lett. **32**, 2266-2268 (2007).
5. G. Temporão, S. Tanzilli, H. Zbinden, and N. Gisin, "Mid-infrared single-photon counting" Opt. Lett. **31**, 1094-1096 (2006).
6. J. Dam, P. Tidemand-Lichtenberg and C. Pedersen, "Room-temperature mid-infrared single-photon spectral imaging" Nat. Photonics. **6**, 788, 2012.
7. H. Xia, G. S., M. Shangguan, X. Xia, X. Jia, C. Wang, J. Zhang, J. S. Pelc, M. M. Fejer, Q. Zhang, X. Dou, and J.-W. Pan, "Long-range micro-pulse aerosol lidar at 1.5 μm with an upconversion single-photon detector," Opt. Lett. **40**, 1579-1582 (2015).
8. L. Høgstedt, A. Fix, M. Wirth, C. Pedersen, and P. Tidemand-Lichtenberg, "Upconversion-based lidar measurements of atmospheric CO_2," Opt. Express **24**, 5152-5161 (2016).

Laser Ranging to Satellites and Space Debris

Georg Kirchner, Michael Steindorfer, Peiyuan Wang, Franz Koidl
Austrian Academy of Science, Space Research Institute

More than 40 Satellite Laser Ranging (SLR) stations around the world routinely measure distances to about 150 retro-reflector-equipped satellites, from Low-Earth-Orbiters (LEO), up to the satellites in geostationary orbit, and determine their orbits with an accuracy of down to few millimetres. Lasers with 10 Hz up to several kHz repetition rate are used, mostly emitting at 532 nm wavelength, and with energies from a few μJ up to several 10 mJ per shot. Most SLR stations are able to detect single photons, reflected from the targets in space, even in full daylight conditions.

SLR already started in the 1960s, quite short after the first success in building an operational laser; with continuous improvements until now. This technique now is also used to range to space debris targets, like old rocket bodies, defunct satellites etc; because of their large number (> 30.000, tracked routinely by large radar systems), they are an increasing problem, and start to endanger present and future space activities. Because these bodies do not have any retro-reflectors, laser ranging has to rely here on diffuse reflection, which requires stronger lasers, and limits achieved ranging distances to about 2000 km – which however covers the most heavily populated LEO orbits. Such space debris laser ranging data allows big improvements – from several 100 m to few 10 m accuracy - for orbit determination for the next few days, assisting in decisions for possibly necessary collision avoidance manoeuvres.

Using the same single-photon detectors, it is also possible to count the flux of single sun photons reflected from space debris targets into the SLR telescopes. This allows to record high-resolution light curves, showing many details of any spin parameters.

Future developments are targeting very high laser repetition rates (up to MHz), fully automatic SLR stations, ranging at near-IR wavelengths, and setting up of a dedicated space debris laser ranging network.

Laser-based Time Transfer through Free-space Links

Ivan Prochazka,
Josef Blazej, Tereza Flekova, Jan Kodet
Czech Technical University in Prague, Brehova 7, 115 19 Prague 1, Czech Republic

The two-way time transfer is an effective way to synchronize two independent time scales with high precision and accuracy independently on the variations of the interconnecting channel [1]. We are reporting on a new approach to an optical two-way free space time transfer which is based on signals of individual photons. This approach enables to reach extreme timing stabilities and minimal systematic errors using existing electro-optic technologies. In our previous work we have demonstrated electronic circuits for two-way time transfer via a single coaxial cable with picosecond accuracy and precision [2]. We have designed and tested the optical analogy of the two-way time transfer using a common optical channel. Photon counting approach to the signal detection is providing several key advantages: the reduction of most systematic errors found in commonly used multi-photon detection systems and the capability to operate with ultimately low signals. The repetitive optical signals of an average intensity as low as 1×10^{-4} photon per pulse may be detected and time tagged with sub-picosecond precision and stability [3]. The principle is illustrated in figure 1.

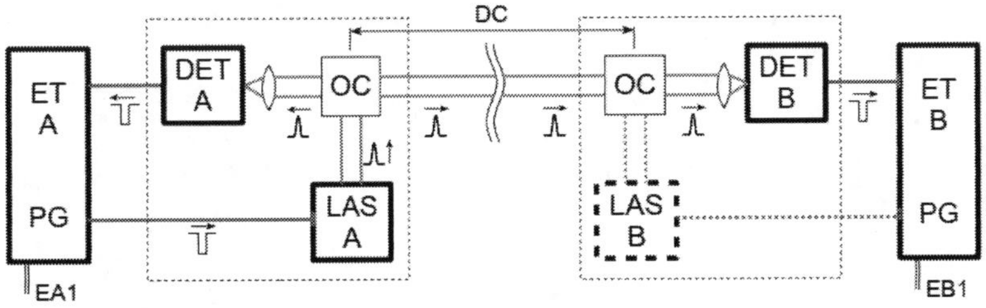

Fig. 1 Block scheme of optical two-way time transfer. Red lines stand for electrical signal lines, green lines stand for optical signal paths. The first measurement phase when module *A* is generating optical signals is displayed.

Two event timers *ET A* and *ET B* are equipped with identical optical modules. The event timers represent the time scales *A* and *B*. Each optical module consists of an optical signal combiner *OC*, laser optical source *LAS* and a photon counting detector *DET*. These modules are interconnected by an optical channel: free space in our case. The measurement is carried out in two phases. First the laser *LAS A* is generating an ultrashort optical pulse. It is split by a signal combiner *OC A* and is propagating in both directions toward *DET A* and *DET B*. They are detected by these detectors and time tagged in the respective time scales *A* and *B*. The corresponding epochs are recorded and stored. In the second phase the laser *LAS B* is generating ultrashort uniform pulse, which is split and propagating toward both detectors *DET A* and *DET B*. The corresponding epochs are recorded. Two pairs of epochs can be combined producing the time difference *DS* between the two time scales along with the delay of the optical channel *DC* interconnecting the two optical modules. Reference points are defined symmetrically inside each optical signal combiner with picosecond accuracy.

The performance and capabilities of described scheme was demonstrated in a number of ground experiments. The limiting precision was better than 400 fs, the long term stability was better than 1 ps over days of operation. The accuracy of optical two way time transfer better than 3 ps was achieved. Analogical two way time transfer was completed via optical fiber which replaced the free space signal propagation. The timing performance achieved was similar to a free space one.

Laser time transfer based on photon counting approach is used also for ground to space time scales comparison. Such a scheme is an extension of the existing satellite laser ranging. Its great advantage is that it completely compensates all the influence of Earth atmosphere and troposphere delays down to sub-picosecond level. Several space missions based on photon counting laser time transfer were successfully completed last years. Recently European Space Agency [4] and China are preparing two space mission with laser time transfer ground to space comparison of atomic and optical clocks on ground and onboard space stations.

References

[1] J. Levine, "A review of time and frequency transfer methods," Metrologia **45**, S162-S174 (2008).
[2] P. Pánek, J. Kodet, and I. Procházka, "Accuracy of two-way time transfer via a single coaxial cable," Metrologia **50**, 60-65 (2013).
[3] I. Procházka, J. Kodet, and J. Blažej, "Note: Solid state photon counters with sub-picosecond timing stability," Review of Scientific Instruments **84**, 046107 (2013).
[4] M. P. Heß, L. Stringhetti, B. Hummelsberger, K. Hausner, R. Stalford, et al., "The ACES mission: system development and test status," Acta Astronautica **69**, 929 (2011).

Yb:CALGO bulk oscillator generating ultrashort pulses at high efficiency by cross-polarized optical pumping

François Labaye[1], Valentin J. Wittwer[1], Norbert Modsching[1], Olga Razskazovskaya[1], Eric Cormier[2,3], Thomas Südmeyer[1]

1. Laboratoire Temps-Fréquence, Institut de Physique, Université de Neuchâtel, Avenue de Bellevaux 51, 2000 Neuchâtel, Switzerland
2. Laboratoire Photonique, Numérique et Nanosciences, UMR 5298, CNRS-IOGS-Université Bordeaux, 33400 Talence, France
3. Institut Universitaire de France (IUF), 1 rue Descartes,75231 Paris

Numerous applications rely on transform-limited few-cycle laser pulses and are typically based on green-pumped Ti:sapphire laser oscillators [1,2]. Simple and cost-efficient Yb-doped bulk laser oscillators are a promising alternative, but have so far been severely limited in power and efficiency for operation in the sub-30-fs regime [3,4]. Standard colinearly-pumped Yb-doped bulk laser oscillators use dichroic pump mirrors with a high transmission for the pump at around 980 nm and a high reflectivity at the lasing wavelengths [Fig. 1a)]. This leads to a strongly varying group delay dispersion (GDD) and limits the expansion of the optical spectrum towards short wavelengths [5]. Furthermore, to reach higher power levels, higher output coupling is required, resulting in a stronger population inversion. For common broadband Yb-doped gain materials, this shifts the gain maximum towards shorter wavelengths until reaching the spectral edge of the dichroic mirrors. Both reasons set a severe limitation to fully exploit the available gain bandwidth for the gereation of even shorter pulses and hinder further power scaling of Yb-doped bulk oscillators in the sub-30-fs regime.

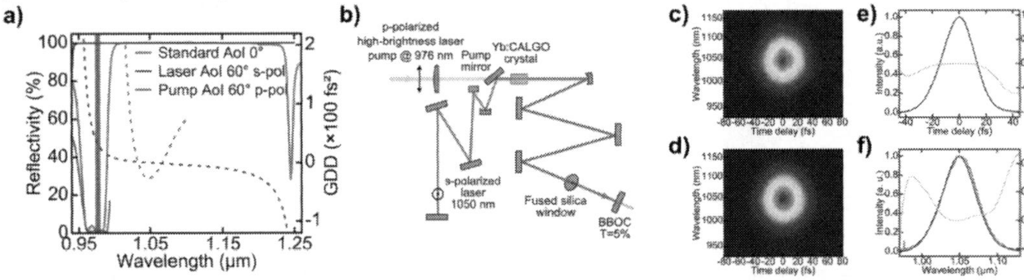

Fig. 1 a) Typical reflectivity (solid line) and GDD (dashed line) of a standard dichroic mirror (red) in comparison to the developed novel pump mirror (blue), pump wavelength at 976 nm (violet line). b) Schematic of the laser oscillator. c-d) Measured and retrieved SHG-FROG trace. e) Retrieved pulse intensity profile. f) Optical spectrum of the generated pulses and retrieved spectral phase.

Here, we show that we can overcome these spectral and power limitations by using a novel pump mirror design relying on cross-polarized optical pumping. The optical coatings utilized in our experimental setup [Fig. 1b)] have been designed and manufactured by ion-beam-sputtering in our facility. The developed pump mirror is highly transmissive for the pump light in p-polarization, while it is highly reflective for the laser light in s-polarization, even covering the pump wavelength and beyond [Fig. 1a)]. In contrast to a standard thin-film polarizer, it provides a low and flat GDD over a broad spectral range for the laser polarization. Applying this pumping scheme to a soft-aperture Kerr-lens mode-locked Yb:CALGO bulk laser oscillator, we achieve 0.5 W of average output power in 28-fs pulses with an optical-to-optical efficiency of 15%. The optical spectrum has a FWHM bandwidth of 55 nm that already starts to extend even beyond the pump wavelength. A SHG-FROG measurement confirms the generation of almost ideal transform-limited soliton pulses [Fig. 1c-f)]. By increasing the OC rate from 5% to 18%, we achieved 1.5 W of average power with slightly longer 31-fs pulses at an optical-to-optical efficiency of 33%.

We overcome the power and efficiency limitations of colinearly-pumped Yb-doped bulk oscillators utilizing a novel cross-polarized optical pumping scheme. Applied to an Yb:CALGO bulk oscillator, we demonstrate up to one order of magnitude higher average power and efficiency compared to other Yb-doped bulk oscillators in the same pulse duration regime [3–5]. As we did not encounter any thermal issues in the gain medium, we expect this approach to enable efficient Yb-doped bulk oscillators operating at the multi-Watt-level in the sub-30-fs regime delivering clean soliton pulses.

References

[1] U. Morgner, F. X. Kärtner, S. H. Cho, Y. Chen, H. A. Haus, J. G. Fujimoto, E. P. Ippen, V. Scheuer, G. Angelow, et T. Tschudi, "Sub-two-cycle pulses from a Kerr-lens mode-locked Ti:sapphire laser," Optics Letters **24**(6), 411 (1999).
[2] D. H. Sutter, G. Steinmeyer, L. Gallmann, N. Matuschek, F. Morier-Genoud, U. Keller, V. Scheuer, G. Angelow, et T. Tschudi, "Semiconductor saturable-absorber mirror–assisted Kerr-lens mode-locked Ti:sapphire laser producing pulses in the two-cycle regime," Optics Letters **24**(9), 631 (1999).
[3] J. Ma, X. Xu, D. Shen, et D. Tang, "21-fs Kerr-lens Mode-locked Yb:CaYAlO4 Laser," in *Conference on Lasers and Electro-Optics (2018)* (Optical Society of America, 2018), paper JTh2A.158.
[4] S. Kimura, S. Tani, et Y. Kobayashi, "Raman-assisted broadband mode-locked laser," Scientific Reports **9**(1)(2019).
[5] P. Sévillano, P. Georges, F. Druon, D. Descamps, et E. Cormier, "32-fs Kerr-lens mode-locked Yb:CaGdAlO$_4$ oscillator optically pumped by a bright fiber laser," Optics Letters **39**(20), 6001-6004 (2014).

https://doi.org/10.1051/epjconf/202024310002

33 W OPCPA at 10 kHz repetition rate with four cycle pulse duration at 2.1 µm based on a single pump laser

Anke Heilmann[1], Martin Bock[1], Lutz Ehrentraut[1], Stefan Eisebitt[1], and Matthias Schnürer[1]
1. Max-Born-Institut, Max Born Straße 2a, 12489 Berlin, Germany

Powerful Mid-IR sources are key tools for advanced spectroscopic applications, high field science and conversion to high frequency radiation. In particular, the so-called water window spectral region between 2.3 nm and 4.4 nm attracts a lot of attention, as it covers the K-shell absorption edges of carbon, nitrogen and oxygen and is therefore ideally suited for spectroscopic studies of biomolecules in their natural aqueous environment [1].

On a laboratory scale, pulses with a central wavelength of a few nm can be obtained by generating high harmonics, where the physics of this highly nonlinear process favors the use of driver lasers in the 2 µm wavelength region. However, due to the extremely low conversion efficiency, meaningful soft X-ray photon flux can only be provided by employing high average-power high pulse energy mid-infrared driver lasers [2].

Here we present a 2.1 µm OPCPA delivering 3.3 mJ pulses at a repetition rate of 10 kHz, thus improving our previous results [3] and setting a new record in terms of average power in the 2 µm wavelength region.

Our setup starts with a commercially available 500 W Yb:YAG thin disk laser (DIRA 500, Trumpf Scientific Lasers) operating at 10 kHz, serving both as pump and signal generation source. The latter is performed in a front-end developed by Fastlite, and makes use of supercontinuum generation, followed by difference frequency generation. The resulting pulses are centered around 2.1 µm and pre-amplified to an average power of 280 mW.

Owing to the system design, the 2 µm pulses should feature passive carrier-envelope phase (CEP) stability. Indeed, first measurements of the single shot CEP noise in our setup yielded results around 200 mrad. However, further improvements are expected since in comparable systems values below 100 mrad have been reported [4].

For power amplification, the pulses are stretched to 1 ps duration with a pair of silicon wedges, so that they fit the pump pulse duration of 2 ps in the two power OPA stages. The first one consists of a 2 mm thick BiBO crystal pumped with 130 W of the DIRA output power. This allows to amplify the 2 µm beam to an average power of 10 W. The second OPA stage is a 5 mm thick YCOB crystal pumped with 320 W, and delivers an output power of 33 W. The amplified pulses are compressed with a 4 cm thick suprasil glass block.

We obtain a slightly elliptic beam profile, as shown in Fig. 3(a). The measured spectrum (Fig.3(b)) extends from 1.8 µm to 2.4 µm, with a FWHM of 430 nm. The modulations result from our front-end configuration, which is optimized for the shortest pulses after power amplification. The spectrum would allow for a Fourier transform limited pulse duration of only 22 fs, which is close to the experimental value of 25 fs inferred from an autocorrelation measurement (Fig. 3(c)).

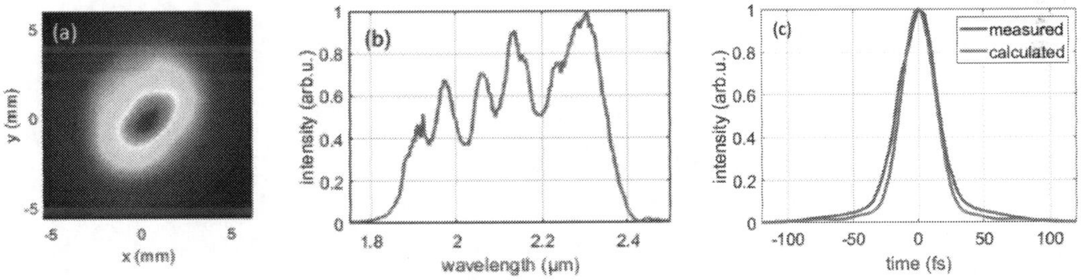

Fig. 1 Characterization of our power OPCPA system: (a) Near field spatial beam profile, (b) measured spectrum (FWHM = 430 nm) and (c) autocorrelation function, where the measured curve is shown in blue (FWHM = 34 fs) and the one calculated from the measured spectrum is shown in orange. The inferred pulse duration is 25 fs.

With these unique parameters our setup is well suited for future experiments in strong field and attosecond physics including the generation of high harmonics in the water window. Since the system will serve as an application laboratory, further development will be dedicated to improve the stability and reliability of its output.

References

[1] Xiaoming Ren *et. al.*, "Attosecond light sources in the water window", J. Opt. **20**, 023001 (2018)
[2] Tenio Popmintchev *et.al.*, "Bright coherent ultrahigh harmonics in the keV x-Ray regime from mid-infrared femtosecond lasers", Science **336**, 1287-1291 (2012)
[3] Tianli Feng *et. al.*, "27 W 2.1 µm OPCPA system for coherent soft X-ray generation operating at 10 kHz," Opt. Express **28**, 8724-8733 (2020)
[4] Nicolas Thiré *et. al.*, "Ultra-stable OPCPA at 2 µm, 16 fs, sub 100 mrad CEP noise", in *High-brightness Sources and Light-driven Interactions Congress*, (Optical Society of America, 2020), p. MM3C.3

The C-cavity, a highly versatile and simple laser design

Robert Lindberg[1], Fredrik Laurell[1], Krister Fröjdh[2] and Walter Margulis[3]
1. Department of Applied Physics, Royal Institute of Technology, 10691 Stockholm, Sweden
2. Proximion AB, Skalholtsgatan 10, SE 164 40 Kista, Sweden
3. Department of Fiber Optics, RISE Acreo, 164 40 Kista, Sweden
Author e-mail address: rl@laserphysics.kth.se

Electrically wavelength-tunable lasers employing chirped fiber Bragg gratings (CFBG) date back to 1993 [1], when a laser diode was connected to a CFBG and operated with a modulated driving current. Varying the modulation frequency thus enabled feedback from different regions of the CFBG, and hence wavelength-tuned operation. This linear cavity design was complemented by a ring variation, oftentimes referred to as a sigma-cavity, five years later [2] in an Er-fiber laser. The sigma cavity configuration has since also been used in lasers employing semiconductor optical amplifiers (SOA) as a gain medium as well as a modulator [3].

As the wavelength tuning in both of these designs is accompanied by varying repetition rates, they are ill-suited for applications relying on synchronization with other sources. It was not until 2017 that a wavelength-tunable design with a constant repetition rate was demonstrated [4]. This design used a fiber Bragg grating array (FBGA) instead of a CFBG to realize stepped tuning, and revolved around reflecting the circulating light from both ends of the FBGA. As the design was unidirectional, this was realized by connecting the FBGA to the mid-ports of two circulators, thus equalizing the overall path through the FBGA, that fromed a ring cavity.

We present a simplified version of this design by employing a linear cavity configuration, which eliminates the need for two circulators [5]. Our laser is realized with a SOA, working as a gain medium and a modulator, and is schematically depicted in Fig. 1 a). The electrical gating sequence, consisting of a repeating pulse pair, that drives the SOA is shown in Fig. 1 b). The separations between these pulses correspond to the time it takes the light to exit the SOA, get reflected from the CFBG and return to the SOA in each arm –as indicated by the color coding. This gating sequence thus inhibits circular propagation, and we refer to the configuration as a C-cavity -due to the resemblance to a "bent" linear cavity. Tuning is thus realized by simply varying the pulse separation by different fractions of the overall round-trip time. The tuning characteristic of this laser is presented in Fig. 1 c), from which it is seen that the tuning range extended 35 nm –limited by the CFBG's reflection bandwidth. The linewidths were about 2 nm, which could be reduced by using shorter gate pulses and/or a more strongly chirped grating. It should also be mentioned that configurations employing an FBGA instead of a CFBG results in linewidths primarily limited by the gratings' reflection bandwidths.

Despite the simple cavity configuration, this design enables highly versatile and tailorable means of operation by interleaving gating sequences. This not only enables average power scaling, but also time-multiplexed multi-wavelength operation as illustrated in Fig. 1 d). The gating sequences can also employ a common gate in one arm, which results in temporally split and overlapping optical pulses at the two outputs –as shown in Fig. 1 e). This immense operation flexibility makes the C-cavity design adaptable to a multitude of applications.

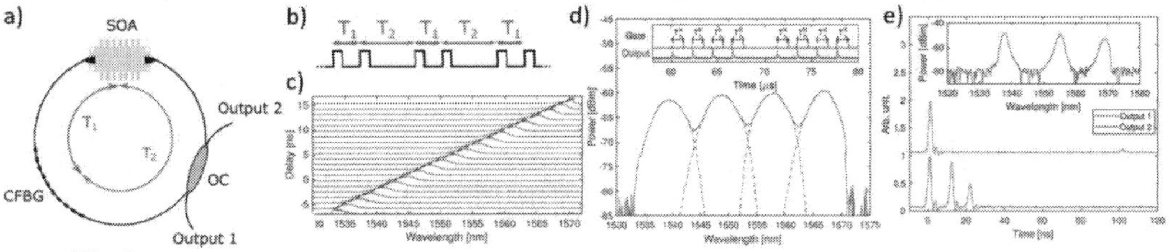

Fig. 1 A schematic depiction of the C-cavity layout is given in a) and the gating sequence shown in b). The achieved tuning characteristics are shown in c). Time-multiplexed four wavelength operation is demonstrated in d) and e) shows the output pulse sequences from the two outputs when using a common gate in one arm.

References

[1] P. A. Morton et al, "Mode-Locked Hybrid Soliton Pulse Source with Extremely Wide Operating Frequency Range," IEEE Photonics Technol. Lett. 5(1), 28–31 (1993).
[2] S. Li and K. T. Chan, "Electrical Wavelength-Tunable Actively Mode-Locked Fiber Ring Laser with a Linearly Chirped Fiber Bragg Grating," IEEE Photonics Technol. Lett. 10(6), 799–801 (1998).
[3] X. Yang el al, "Continuously tunable, narrow-linewidth laser based on a semiconductor optical amplifier and a linearly chirped fiber Bragg grating," Opt. Express 27, 14213-14220 (2019)
[4] T. Tiess, M. Becker, M. Rothhardt, H. Bartelt, and M. Jäger, "Discrete tuning concept for fiber-integrated lasers based on tailored FBG arrays and a theta cavity layout," Opt. Lett. 42(6), 1125–1128 (2017).
[5] R. Lindberg et al., "C-cavity fiber laser employing a chirped fiber Bragg grating for electrically gated wavelength tuning," Opt. Express 28, 9208-9215 (2020)

EPJ Web of Conferences **243**, 11002 (2020) https://doi.org/10.1051/epjconf/202024311002
EUROPHOTON 2020

Using the Variable Pump Intensity method to measure optical gains and unveil photophysical and photonic phenomena in active waveguides

Luis Cerdán,[1]* Marco Anni,[2] Maria Luisa De Giorgi,[2] Pedro G. Boj,[3] María Ángeles Díaz-García[4]

1. Instituto de Química Física "Rocasolano", Consejo Superior de Investigaciones Científicas (CSIC), C/ Serrano 119, 28006, Madrid, Spain (Affiliation until 2017). e-mail: lcerdanphd@gmail.com

2. Dipartimento di Matematica e Fisica "Ennio de Giorgi", Università del Salento, via per Arnesano, 73100, Lecce, Italy

3. Departamento de Óptica, Farmacología y Anatomía, Instituto Universitario de Materiales de Alicante y Unidad Asociada UA-CSIC, Universidad de Alicante, 03080, Alicante, Spain

4. Departamento de Física Aplicada, Instituto Universitario de Materiales de Alicante y Unidad Asociada UA-CSIC, Universidad de Alicante, 03080, Alicante, Spain

The last decades have witnessed the advent of a myriad of new solution-processed laser materials that can be implemented as low-cost, flexible, and/or integrated waveguide devices [1]. This profusion makes necessary the development of reliable methods to delineate all their amplifying signatures and thus to open the door to appropriate cross-sample comparisons. In this sense, the generation of Amplified Spontaneous Emission (ASE) is routinely used to assess the performance of active materials prior to developing thin film lasers. In particular, the Variable Stripe Length (VSL) method has been profusely used for decades to measure the optical gain in these materials and devices, but experimental and theoretical evidences cast doubts upon its reliability [2]. Seeking to solve those problems, a new formalism to fully characterize the amplification properties of active waveguides based on the most common Variable Pump Intensity (VPI) method has been recently reported by one of us [3]. The analytical expression describing the growth of ASE intensity $I_{ASE}(\lambda, I_p, L)$ as a function of the wavelength, pump value (I_p; energy/density/intensity/flux/…) and stripe length (L) reads [3]:

$$I_{ASE}(\lambda, I_p, L) = \frac{\Omega(\lambda)\eta(I_p)}{\gamma(\lambda)\Delta n(I_p) - \alpha(\lambda)} \left(e^{(\gamma(\lambda)\Delta n(I_p) - \alpha(\lambda))L} - 1\right), \tag{1}$$

with the population inversion Δn following a well-defined but cumbersome rational function. It was shown that eq. (1) allows retrieving from the ASE spectra of a single VPI experiment, the spectrally resolved losses coefficient $\alpha(\lambda)$ and net optical gains $g = \gamma\Delta n - \alpha(\lambda)$ at all pump values (Fig. 1).

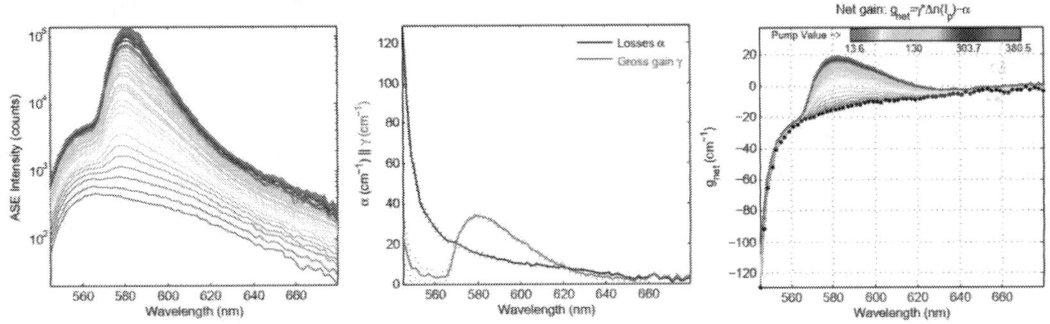

Fig. 1 (Left panel) ASE spectra as a function of pump intensity used for the fit. (Centre panel) Fitted losses α and gross gain γ as a function of wavelength. (Right panel) Spectrally resolved net optical gain at increasing pump intensities. Sample: 1 µm thick 10 mM Rhodamine 6G (Rh6G) doped pHEMA film deposited onto a quartz substrate.

In this communication, we will revisit the main fundaments of the VPI method and its implementation into the freely distributed software *ASE gain Labs,* and will highlight the potential of this methodology to unveil relevant information on the photonic properties of the waveguiding devices (mode confinement effects, leaky-modes, …) and on the photophysics of the active materials (excited state absorption, aggregates, …) [4]. In other words, we will show that the VPI method is a relevant, time-saver, and powerful formalism.

References

[1] M. Anni, S. Lattante (Eds.), "Organic Lasers: Fundamentals, Developments, and Applications" (Pan Stanford Publ., Singapore, 2018)
[2] L. Cerdán, "Variable Stripe Length method: influence of stripe length choice on measured optical gain," Opt. Lett. **42**, 5258-5261 (2017)
[3] L. Cerdán, "Simultaneous retrieval of optical gains, losses, and threshold in active waveguides," Opt. Laser Technol. **121**, 105814 (2020)
[4] L. Cerdán, M. Anni, M. L. De Giorgi, P. G. Boj, M. A. Díaz-García, "Probing the limits of the Variable Pump Intensity method to measure optical gains and unveil photophysical and photonic phenomena in waveguides and solutions," submitted.

Comparative study of Ho:$Y_3Al_5O_{12}$ and Ho:Y_2O_3 transparent ceramics synthesized from laser ablated nanopowders

Liza Basyrova[1,*], Roman Maksimov[2,3], Vladislav Shitov[2], Alexander Khubetsov[4], Olga Dymshits[4], Mikhail Baranov[1], Florent Starecki[5], Patrice Camy[5], and Pavel Loiko[5]

*1. ITMO University, 49 Kronverkskiy Pr., 197101 St. Petersburg, Russia. *E-mail: lizaveta.lel@gmail.com*
2. Institute of Electrophysics, Ural Branch of the Russian Academy of Sciences, 106 Amundsen St., 620016 Ekaterinburg, Russia
3. Ural Federal University named after the first President of Russia B.N. Yeltsin, 19 Mira St., 620002 Ekaterinburg, Russia
4. Vavilov State Optical Institute, 36 Babushkina St., 192171 St. Petersburg, Russia
5. Centre de recherche sur les Ions, les Matériaux et la Photonique (CIMAP), UMR 6252 CEA-CNRS-ENSICAEN, Université de Caen Normandie, 6 Boulevard du Maréchal Juin, 14050 Caen Cedex 4, France

Holmium (Ho^{3+}) ions are known for their emission at 2 μm originating from the $^5I_7 \rightarrow {}^5I_8$ transition. Ho lasers find applications in coherent laser radar, atmospheric sensing, surgery and frequency conversion into mid-IR. Among the laser host crystals for Ho^{3+} doping, garnets ($Y_3Al_5O_{12}$) and sesquioxides (A_2O_3, A = Lu, Y, Sc) have attracted attention due to their attractive thermal and spectroscopic properties. Transparent polycrystalline ceramics offer the advantages of lower synthesis temperatures, ease of doping and size-scalable production [1]. We report on a comparative study of Ho:$Y_3Al_5O_{12}$ and Ho:Y_2O_3 transparent ceramics promising for thin-disk ~2 μm lasers.

The 0.94 at.% Ho:$Y_3Al_5O_{12}$ ceramic was fabricated by vacuum sintering at 1780 °C for 20 h using 0.5 wt% TEOS as a sintering aid from Ho:Y_2O_3 and Al_2O_3 nanopowders produced by laser ablation and sedimentation [2]. For the 0.94 at.% Ho:Y_2O_3 ceramics, we used laser-ablated Ho:Y_2O_3 nanopowder with an addition of ~2 wt% ZrO_2 as a sintering aid to avoid exaggerated grain growth. They were uniaxially pressed and vacuum sintered at 1780 °C for 20 h. The obtained ceramics were transparent with yellowish coloration, Fig. 1(a).

The ceramics were of single-phase nature. The Ho:$Y_3Al_5O_{12}$ one exhibited a cubic structure (sp. gr. $Ia\bar{3}d$, a = 10.010 Å), Fig. 2(a), and a mean grain size D_{grain} of 32 μm. The Ho:Y_2O_3 ceramics was also cubic (sp. gr. $Ia\bar{3}$, bixbyite (C-type) structure, a = 10.569 Å) with smaller D_{grain} of 6.8 μm. Both ceramics exhibited a close-packed microstructure with clean grain boundaries and a lack of pores, Fig. 1(c).

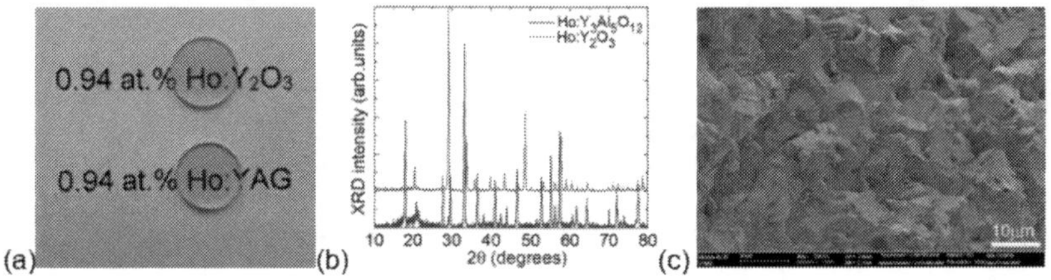

Fig. 1 (a) Photograph of the laser-grade polished ceramic disks; (b) X-ray powder diffraction (XRD) patterns of ceramics; (c) Scanning Electron Microscope (SEM) image of the fracture surface of the Ho:Y_2O_3 ceramics.

The ceramics exhibited high transmission in the visible and IR, Fig. 2(a). For the Ho:Y_2O_3 ceramics at 1.0 μm, T_0 = 80.0% which is close to the theoretical limit set by Fresnel losses, T_{Fr} = 81.6%. The transition intensities were analysed within the Judd-Ofelt theory. Absorption, σ_{abs}, and stimulated-emission (SE), σ_{SE}, cross-sections, Fig. 2(b), for the $^5I_7 \leftrightarrow {}^5I_8$ Ho^{3+} transition were determined and the luminescence lifetimes of several Ho^{3+} excited-states were measured. The multi-phonon non-radiative (NR) relaxation rates were estimated. For the Ho:$Y_3Al_5O_{12}$ ceramics, $\tau_{lum}(^5I_7)$ = 6.85 ms, Fig. 2(c), which is close to the estimated radiative lifetime (7.82 ms).

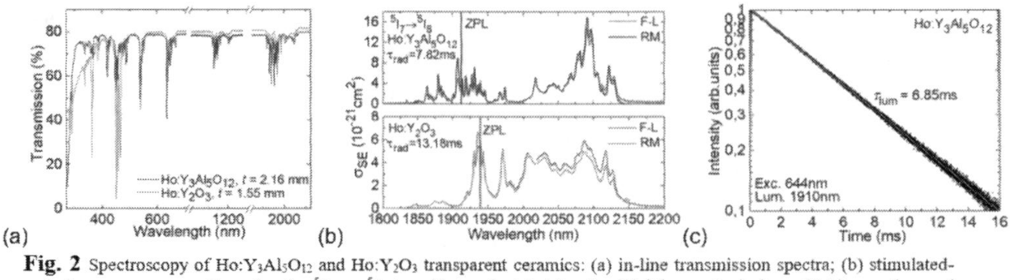

Fig. 2 Spectroscopy of Ho:$Y_3Al_5O_{12}$ and Ho:Y_2O_3 transparent ceramics: (a) in-line transmission spectra; (b) stimulated-emission, σ_{SE}, cross-sections for the $^5I_7 \rightarrow {}^5I_8$ transition calculated by the Füchtbauer–Ladenburg (F-L) equation and reciprocity method (RM); (c) luminescence decay curve from the $^5I_7$ Ho^{3+} state for the Ho:$Y_3Al_5O_{12}$ ceramics.

References

[1] H. Yang, J. Zhang, X. Qin, D. Luo, J. Ma, D. Tang, H. Chen, D. Shen, and Q. Zhang, "Polycrystalline Ho:YAG transparent ceramics for eye-safe solid state laser applications," J. Am. Ceram. Soc. **95**(1), 52-55 (2012).

[2] L. Basyrova, R. Maksimov, V. Shitov, M. Baranov, V. Mikhaylovsky, A. Khubetsov, O. Dymshits, X. Mateos, and P. Loiko, "Effect of SiO_2 addition on structural and optical properties of Yb:$Lu_3Al_5O_{12}$ transparent ceramics based on laser ablated nanopowders," J. Alloy Compd. **806**, 717-725 (2019).

Growth, spectroscopy and diode-pumped laser operation of acentric Yb:KGd(PO₃)₄ crystal

Shanming Li[1,2], Anna Volokitina[1,3,*], Rosa Maria Solé[1], Pavel Loiko[4], Valentin Petrov[5], Uwe Griebner[5], Yin Hang[2], Francesc Díaz[1], Magdalena Aguiló[1] and Xavier Mateos[1]

1. Universitat Rovira i Virgili (URV), FiCMA-FiCNA-EMaS, Marcel.lí Domingo 1, 43007 Tarragona, Spain
2. Laboratory of Micro-Nano Photonic and Optoelectronic Materials and Devices, Key Laboratory of Materials for High Power Laser, Shanghai Institute of Optics and Fine Mechanics, Chinese Academy of Sciences, 201800 Shanghai, China
*3. ITMO University, 49 Kronverkskiy Pr., 197101 St. Petersburg, Russia. *E-mail: anna.itmo@gmail.com*
4. CIMAP, UMR 6252 CEA-CNRS-ENSICAEN, Université de Caen Normandie, 6 Boulevard du Maréchal Juin, 14050 Caen, France
5. Max-Born-Institute for Nonlinear Optics and Short-Pulse Spectroscopy, 2A Max-Born-Str., 12489 Berlin, Germany

Bifunctional crystals capable of hosting laser-active rare-earth ions and allowing for frequency doubling are of interest for applications in compact visible lasers. This can be realized in crystals exhibiting non-centrosymmetric structure, which provides the necessary $\chi^{(2)}$ nonlinearity. So far, such crystals were mainly doped with Nd^{3+} ions. However, ytterbium (Yb^{3+}) ions are also very attractive for such laser designs as they provide longer upper laser level lifetimes, higher Stokes pump efficiency, weaker heat loading and broader emission bandwidths which is of interest for the generation of fs pulses in the green spectral range [1]. In the present work, we report on the growth, polarized spectroscopy and first diode-pumped laser operation of an acentric crystal, Yb:KGd(PO_3)$_4$.

The Yb:KGd(PO_3)$_4$ crystals were grown by the top-seeded solution growth slow-cooling method, Fig. 1(a). The doping concentration was measured by electron probe microanalysis, $N_{Yb} = 1.0\times10^{20}$ cm^{-3} (2.4 at.% Yb). Yb:KGd(PO_3)$_4$ is monoclinic (sp. gr. $P2_1$) and optically biaxial (it is phase-matchable for type I second harmonic generation) [2]. The absorption, σ_{abs}, and stimulated-emission (SE), σ_{SE}, cross-sections for Yb^{3+} measured for light polarized along the optical indicatrix axes are shown in Fig. 1(b). The maximum σ_{SE} in the region where the laser operation is expected, amounts to 0.73×10^{-20} cm^2 at 1006 nm for $\boldsymbol{E} \parallel N_m$. The radiative lifetime of the $^2F_{5/2}$ multiplet is relatively long, 1.22 ms. The Stark splitting for Yb^{3+} resolved at 6 K is shown in Fig. 1(c).

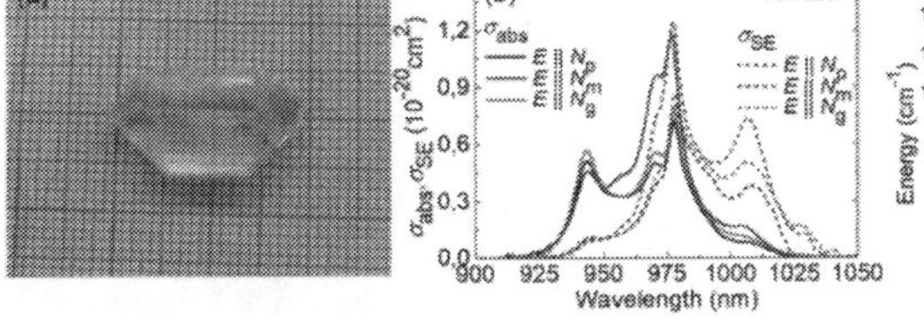

Fig. 1 Growth and spectroscopy of Yb:KGd(PO_3)$_4$: (a) Photograph of the as-grown crystal; (b) absorption, σ_{abs}, and stimulated-emission, σ_{SE}, cross-sections for Yb^{3+} ions with polarized light; (c) crystal-field splitting of Yb^{3+} multiplets.

Laser operation of Yb:KGd(PO_3)$_4$ at the fundamental wavelength (~1 μm) was achieved under diode-pumping by a fiber-coupled InGaAs laser diode at 966 nm. The N_g-cut uncoated crystal was 7.03 mm-thick; it was mounted in a Cu-holder cooled by flowing water (14 °C). A compact plano-plano cavity was employed and the laser mode was stabilized by the positive thermal lens in the crystal. The laser generated a maximum output power of 1.57 W at 1040.7 nm with a slope efficiency of 44.4% (vs. absorbed pump power) and a threshold of 0.68 W (for transmission of the output coupler T_{OC} = 0.5%), Fig. 2(a). The laser emission was linearly polarized ($\boldsymbol{E} \parallel N_m$) and the wavelength experienced a blue-shift with T_{OC}, Fig. 2(b). The laser operated at the fundamental mode, Fig. 2(c).

Yb:KGd(PO_3)$_4$ is promising for mode-locked lasers with self-frequency-doubling yielding green fs pulses.

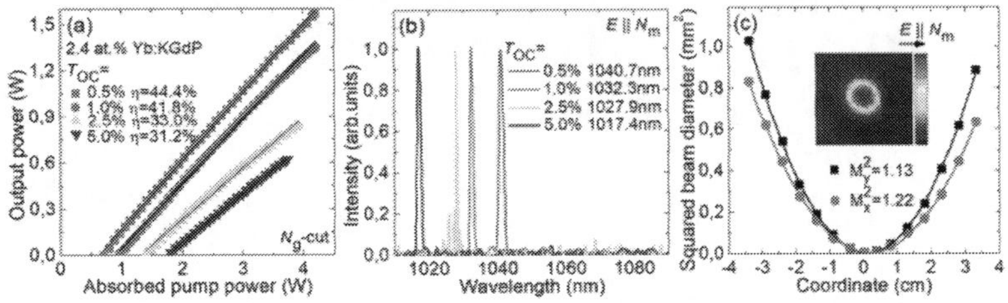

Fig. 2 Diode-pumped 2.4 at.% Yb^{3+}:KGd(PO_3)$_4$ (Yb:KGdP) laser: (a) Input-output dependences, η – slope efficiency; (b) typical spectra; (c) evaluation of the beam quality factors, $M^2_{x,y}$; *inset*: far-field mode profile, T_{OC} = 0.5%, P_{abs} = 3.6 W. The laser polarization is $\boldsymbol{E} \parallel N_m$ (horizontal, $\parallel x$).

References

[1] M. Kowalczyk, X. Zhang, X. Mateos, S. Guo, Z. Wang, X. Xu, P. Loiko, J. E. Bae, F. Rotermund, J. Sotor, U. Griebner, and V. Petrov, "Graphene and SESAM mode-locked Yb:CNGS lasers with self-frequency doubling properties," Opt. Express **27**(2), 590-596 (2019).
[2] I. Parreu, M. C. Pujol, M. Aguiló, F. Díaz, X. Mateos, and V. Petrov, "Growth, spectroscopy and laser operation of Yb:KGd(PO₃)₄ single crystals," Opt. Express **15**(5), 2360-2368 (2007).

EPJ Web of Conferences **243**, 13001 (2020)
EUROPHOTON 2020 https://doi.org/10.1051/epjconf/202024313001

Full-energy, vacuum-compatible, single-shot pulse characterization method for petawatt-level ultra-broad bandwidth lasers using spatial sampling

B. Webb, S. W. Bahk, I. A. Begishev, C. Dorrer, C. Feng, C. Jeon, M. Spilatro, R. Roides, J. Zuegel, and J. Bromage
Laboratory for Laser Energetics, University of Rochester, 250 East River Road, Rochester, NY, 14623-1299, USA

Optical parametric chirped-pulse amplification (OPCPA) enables the amplification of ultra-broad bandwidths beyond the petawatt level, creating the potential for shorter pulse durations than are achievable with traditional CPA. Since these shorter pulses are highly sensitive to material dispersion, all-reflective optics are typically used after the pulse compressor to ensure that the pulse characterized by a diagnostic is the same as the pulse interacting with the target. Because beam splitters, leaky mirrors, and other sampling optics introduce chromatic dispersion and nonlinear distortion, sampling the beam to measure the pulse duration on target at full energy is problematic.

In this work we investigate the accuracy of spatially sampling the beam in order to provide single-shot pulse duration on target at full energy with spectral phase interferometry for direct electric-field reconstruction (SPIDER) [1]. This technique was utilized in the first-light experiments of a new 0.5-PW, 20-fs OPCPA system that was completed in 2020, the Multi-Terawatt-pumped Optical Parametric Amplifier Line (MTW-OPAL) [2]. Precise apodization of a beamlet near the edge of the main 90-mm beam before the compressor using a spatial dithering, allows for a small mirror after the compressor to pick off the beamlet for pulse characterization with only 2.5% loss in overall energy to target (Fig. 1). This method of spatial sampling allows for pulse-shape measurement at the same time the full-energy beam is interacting with the target, providing on-shot pulse shape information for experiments. Vacuum compatibility of a single-shot SPIDER is accomplished by placing spectrometers for the fundamental and upconverted SPIDER signal outside the vacuum chamber. Transmission through a window is permitted after the nonlinear crystal since the spectral phase has already been encoded in the up-converted interferogram.

The local pulse duration is measured at several points across the beam to verify the accuracy of this method. Since this spatially resolved characterization detects changes in dispersion and not group delay, the far-field spot was also characterized to detect the presence of spatiotemporal distortions [3]. Furthermore, the full-aperture beam is characterized at low energy with another SPIDER device and compared to sub-aperture measurement to determine the accuracy of this spatial sampling method. Periodic full-aperture measurements combined with a new spatiotemporal diagnostic [4] can maintain the accuracy of this spatial sampling method over time.

This material is based upon work supported by the Department of Energy National Nuclear Security Administration under Award Number DE-NA0003856, the University of Rochester, and the New York State Energy Research and Development Authority. The support of DOE does not constitute an endorsement by DOE of the views expressed in this abstract.

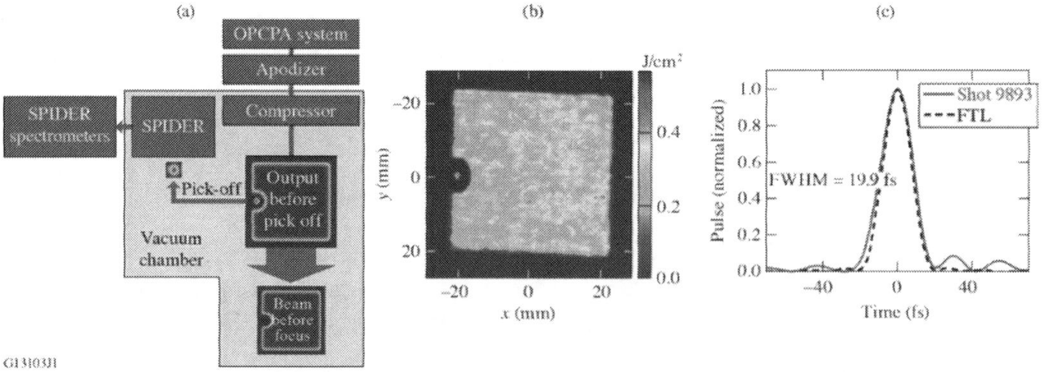

Fig. 1. (a) Schematic of the MTW-OPAL system showing the back end. (b) Output apodized beam profile before the pulse compressor. (c) Pulse calculated from SPIDER measurement compared to the Fourier-transform limit (FTL).

References

[1] C. Dorrer, B. de Beauvoir, C. Le Blanc, S. Ranc, J.-P. Rousseau, P. Rousseau, and J.-P. Chambaret, "Single-Shot Real-Time Characterization of Chirped-Pulse Amplification Systems by Spectral Phase Interferometry for Direct Electric-Field Reconstruction," Opt. Lett. **24**, 1644 (1999).
[2] J. Bromage, S.-W. Bahk, I. A. Begishev, C. Dorrer, M. J. Guardalben, B. N. Hoffman, J. B. Oliver, R. G. Roides, E. M. Schiesser, M. J. Shoup III, M. Spilatro, B. Webb, D. Weiner, and J. D. Zuegel, "Technology Development for Ultraintense All-OPCPA Systems," High Power Laser Sci. Eng. 7, e4 (2019).
[3] C. Dorrer, "Spatiotemporal Metrology of Broadband Optical Pulses," IEEE J. Sel. Top. Quantum Electron. **25**, 3100216 (2019).
[4] S.-W. Bahk, C. Dorrer, R. G. Roides, and J. Bromage, "Chromatic-Aberration Diagnostic Based on a Spectrally Resolved Lateral-Shearing Interferometer," Appl. Opt. **55**, 2413 (2016).

EPJ Web of Conferences **243**, 13002 (2020) https://doi.org/10.1051/epjconf/202024313002
EUROPHOTON 2020

Influences on the direction probabilities for the direction instability phenomenon in fiber ring lasers

M. A. Arshad[1,2]**, *Arni Pratiwi***[1,2]**, *A. Hartung***[1]**, *M. Jäger***[1]

1. Leibniz-Institute of Photonic Technology, Albert-Einstein-Straße 9, 07747 Jena, Germany

2. Friedrich Schiller University, Faculty of Physics and Astronomy, Max-Wien-Platz 1, 07743 Jena, Germany

Recently, we observed a direction instability phenomenon, the spontaneous transition from two emission directions to to one emission direction, in a km-long fiber ring laser [1]. The setup in use is shown in Fig. 1. The cavity is pumped by a pump diode emitting at a wavelength of 976 nm. The pump power enters the ring cavity by a pump signal combiner and reaches a 6 m long Yb-doped active fiber section. Here, without any wavelength selective filter applied, a laser signal at around 1100 nm is emitted. The active fiber is followed by a long (~ km) passive fiber (Corning HI1060). Finally, an output tap coupler with a very low nominal coupling ratio of 0.01% at 1060 nm feeds most of the signal back to the pump signal combiner where it passes through the ring again, and a small part of the signal out of the cavity. As no optical isolator is included in the ring cavity clockwise and counter clockwise signals are allowed to propagate and induced at the laser threshold.

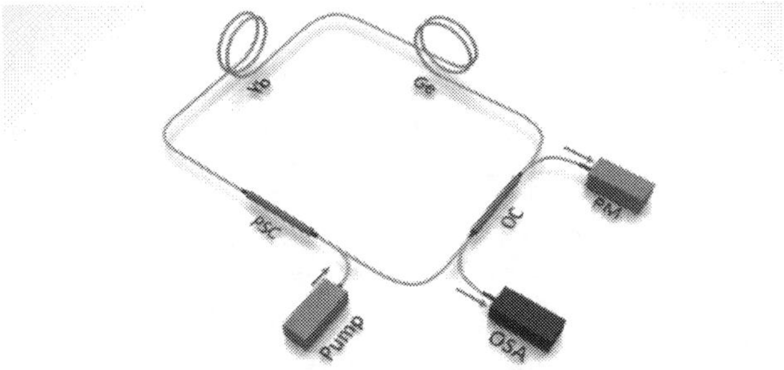

Fig. 1 General scheme of the all-fiber ring laser. The setup consists of a pump diode (pump), pump signal combiner (PSC), an active Yb-doped fiber (Yb), a passive Ge-doped fiber (Ge), and an output coupler (OC).

This cavity shows the effect of spontaneous unidirectional lasing in combination with optical bistability. Unidirectional lasing occurs far above the laser threshold for up to now unknown reasons. The final direction is not predefined. The cavity can finally switch in any of the two directions. We observed that the direction probabilities can be significantly altered by the loss sequence of the passive fiber. We composed the passive fiber by 1 km long sections of two different fibers. Both fibers are Corning's HI 1060 fiber, share the same data sheet, and meet its requirements at the design wavelength of 1060 nm. Nevertheless, the spectral dependence is locally different due to variations in the OH content.

We tested different fiber (and hence loss) sequences of two to four kilometer long cavities. Based on samples of 30 runs for each cavity layout we observed the following trends regarding the direction probabilities. The most equal probabilities were observed for an all-high-loss cavity with 2 km length. Here, one direction was triggered 37% of all runs and the other direction 67% of all runs. In contrast, including the low-loss kilometer seems to enforce a final direction with probabilities up to 100%. In these cases the preferred final direction, starting at the active fiber, passes the high loss fiber first and the low loss fiber thereafter. Flipping the low-loss and the high-loss fiber also flipped the preference of the final direction. Passing the high linear loss fiber first might correspond with reducing the power and hence the nonlinear loss experienced by the laser wavelength. This indicates that nonlinear loss could have a crucial contribution to this new type of instability.

In order to achieve equal probabilities in both directions we set up a component-wise symmetric cavity where the output coupler is centered between two 1 km long identical passive fiber sections. Surprisingly, there is still a strong preference with one final direction being triggered 93% of the times. Only minor differences are left, to explain this imbalance all of them typically considered negligible regarding roundtrip loss like varying splice losses in the range of 0.2 dB. This raises the question how much or how little imbalance is needed to influence the direction probabilities.

References

[1] M. A. Arshad, A. Hartung, and M. Jäger, "A stimulated Stokes Raman scattering-based approach for continuous wave supercontinuum generation in optical fibers," *Laser Phys. Lett.*, vol. 16, no. 3, 2019, doi: 10.1088/1612-202X/aaff53.

EPJ Web of Conferences **243**, 13003 (2020)
EUROPHOTON 2020 https://doi.org/10.1051/epjconf/202024313003

Beam-Shaping Optimization of the diode bar end-pumped laser

Lyuben S. Petrov, Kamen Velev, Kaloyan Georgiev, Ivan Buchvarov
Physics Department, Sofia University, 5 James Bourchier Blvd., BG-1164 Sofia, Bulgaria

Nowadays diode laser bars with theirs compact design, energy scaling by increasing the number of emitters as well as easy adjustment of the wavelength by temperature makes them basic elements for high-power solid-state laser pumping. A single diode bar is a monolithic linear array which can emit near infrared laser radiation from tens up to few hundreds of watts of power in CW or QCW mode, respectively. However the beam profile of a single diode laser bar is highly asymmetric in slow and fast axis and beam-shaping techniques are used to adapted beam-parameter products in both axes [1]. However, at present no easy theoretical procedure has been developed to optimize the operation of the beam-shaping device used for the specific laser bar.

In this paper we proposed a method of optimization of a laser bar beam-shaping device (BSD), using of a model of off-axis multiple Gaussian beam propagation. The method has been used to design and experimental demonstration of the specific beam-shaping device [2]. It chops the incident laser beam into a specific number of beams and then redirects and repositions these beams, so that they emerge from the beam shaper stacked on top of one another. The optimization procedure shows that for each specific bar (the number of emitters, their width and pitch there is an optimum number of the beam "chops". We view each chop in vertical direction (Fast axis) as an individual Gaussian beam and the resulting shaped beam as a sum of offset Gaussian beams. In horizontal direction (Slow axis) the beam is determined by the parameters of the laser diode: emitter width, pitch and number of emitters, which is modified by the number of chops. As such the more chops there are the less Gaussian beams are summed for the Slow axis beam ("number of emitters"/"number of chops").

The propagation of the beam then is calculated using a generalization of the ABCD matrix law which allows for the beam to be off axis [3]. The beam is propagated through a lens and then the beam quality is numerically measured. The evolution of the quality of the resulting shaped beam in the vertical direction (Fast axis) as a function of the number of chops is shown on fig. 1 a), where the ratio, of the pitch (distance peak to peak between two neighboring beams) to the beam width, is varied to show the changes for different beam shaper and laser diode configurations. We have established numerically that if the ratio Pitch/Width is constant so is the M^2.

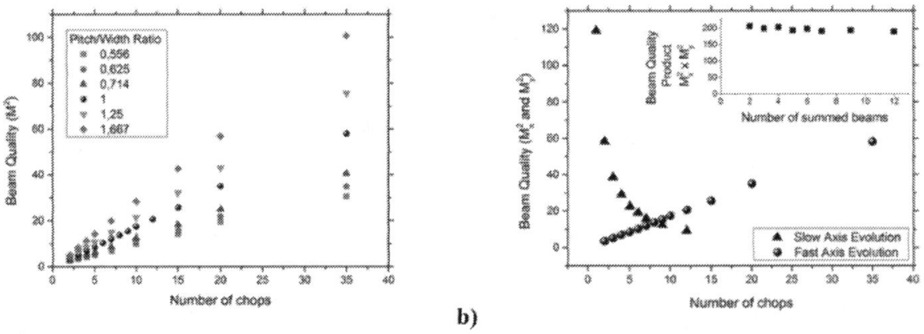

a) **b)**

Fig. 1 a) Shows the evolution of the beam quality with the increase of the number of Gaussian beams summed. b) Shows the M_y^2 (Fast axis) evolution depending on the number of chops (Gaussian beams summed, on the same graph is shown the evolution of the M_x^2 (Slow axis), but the number of chops shows how much of all the emitters are used ("all emitters"/"number of chops"). As an inset the product of both axis Beam Quality is show.

In \Fig.1 b) we show how the optimum amount of chops can be chosen for the beam shaper. The pitch/width ratio we have chosen is 1, under which the losses from the beam shaper increase dramatically. For the laser diode the simulations are based on a Jenoptik JOLD-225-QPFN-1L, who's M_x^2 we have measured and is around 127 and has 37 emitters. The brightness of the beam is proportional to the whole system beam quality which is defined by the product of the M^2 factors of both axes, which is shown as an inset in Fig. 1 b). We can see that it stays more or less the same, so no optimization can be performed there. Based on our simulations the optimum number of chops, to achieve symmetric beam quality for this system, should be around 8.

References

[1] F. M. Dickey and T. E. Lizotte, Laser Beam Shaping Applications, 2nd ed., (CRC Press, 2017)
[2] W. A. Clarkson, D. C. Hanna, "Two-mirror beam-shaping technique for high-power diode bars," Opt. Lett. 21, 375-377 (1996)
[3] A. A. Tovar, L. W. Casperson, "Generalized beam matrices: Gaussian beam propagation in misaligned complex optical systems," J. Opt. Soc. Am. A, Vol. 12, No. 7 (1995)

Nonlinear plasmonic metasurfaces assisted laser mode locking

Lei Zhang,[1,2] Jiyong Wang, [1,2] Aurelien Coillet,[3] Philippe Grelu,[3]*

Benoit Cluzel,[3] Min Qiu[1,2*]*

[1]*Key Laboratory of 3D Micro/Nano Fabrication and Characterization of Zhejiang Province, School of Engineering, Westlake University, 18 Shilongshan Road, 310024 Hangzhou, Zhejiang Province, China*

[2]*Institute of Advanced Technology, Westlake Institute for Advanced Study, 18 Shilongshan Road, 310024 Hangzhou, Zhejiang Province, China*

[3]*Laboratoire Interdisciplinaire Carnot de Bourgogne, Université Bourgogne Franche-Comté, 9 avenue Alain Savary, 21078 Dijon, France*

Abstract

Plasmonic metasurfaces are artificial 2D layers made of subwavelength elementary cells, which give rise to novel wave properties that do not exist in nature. In the linear regime, their applications have been extensively studied, especially in wavefront manipulation for lensing, holography or polarization control. Interests in metasurfaces operating in nonlinear regime have also increased due to their ability to efficiently convert the fundamental light into harmonic frequencies and multiphoton emissions. Nevertheless, practical applications in the nonlinear regime have been rarely reported.

In this study, we report that plasmonic metasurfaces with well-controlled polarimetric nonlinear transfer functions perform as saturable absorbers with modulation performances superior to that of other 2D materials. We employ planar nanotechnologies to fabricate 2D plasmonic metasurfaces with precise size, gap and orientation. We quantify the relationship between saturable absorption and the plasmonic resonances of the unit cell by altering the excitation power of pumping laser, the polarization of incident light and the geometrical parameters of the plasmonic metasurfaces. Finally, we provide a practical implementation by integrating the saturable metasurfaces into a fiber laser cavity and realize a stable self-starting ultrashort laser pulse generation. As such, this work sheds light on ultrathin nonlinear saturable absorbers for applications where nonlinear functions are required, such as in ultrafast laser or neuromorphic circuits.

References:

[1] Wang, J., Coillet, A., Demichel, O. *et al.* "Saturable plasmonic metasurfaces for laser mode locking," *Light Sci Appl* **9**, 50 (2020).

[2] Sain, B., Zentgraf, T. "Metasurfaces help lasers to mode-lock," *Light Sci Appl* **9**, 67 (2020).

CW Nd:YVO$_4$ disk laser with multipoint diode pumping and in-phase super-mode lasing

***D. A. Guryev*[1], *D. A. Nikolaev*[1], *V. B. Tsvetkov*[1]**

1. Prokhorov General Physics Institute, Russian Academy of Sciences, 38 Vavilov Str., 119991 Moscow, Russian Federation

Development of large aperture lasers with diffraction limited output is a relevant problem of laser physics. The use of the degenerated type cavity is a good way to solve this problem. In the degenerated cavity all pump areas can take part in the in-phase collective lasing of a diffraction limited output beam. In present work we demonstrate in-phase super-mode lasing regime in the solid-state disk laser with multipoint diode pumping and results of research of its lasing conditions.

The laser cavity was formed by a flat HR mirror placed on the back surface of an active disk (AD) and a flat output mirror (both coated at λ=1064 lasing wavelengths). The antireflection coating has been covered on forward surface of the AD for the pump (808 nm) and lasing wavelengths. the AD was a 20×20×0.5 mm^3 Nd:YVO$_4$ crystal. The continious-wave laser diode was used to pump the AD. Pump radiation inside the AD was a one-dimensional spatially-periodic pumping points pattern and consisted of 9 spots with the center distance d=0.5mm.

Transverse distributions of laser radiation in a far-field zone and inside the AD were researched in dependence on absorbed pump power P_{in} (0 – 4 W) and cavity length L (6 – 19 cm).

The results of the research showed that laser output had a spatial-periodic structure both in a far-field zone and inside the AD (Figure 1 (a) and (b) respectively) in a range of cavity length values L = 6-11 cm with all pump power levels and in a range of cavity length values L = 12-19 cm with absorbed pump power lower than 3W. Maximum output power was 250 mW. A similar picture also had been observed and explained in paper [1]. In the field of cavity lengths L=12-19 cm and P_{in}=3-4 W the lasing appeared at N=5-7 neighboring pumping points. The transverse intensity distribution of the laser emission inside the AD (Figure 1 (c)) did not possess the spatially periodic character in contrast with the results for another area. The angular structure of radiation in a far-field zone represented an intensive central peak and two weak petals (Figure 1 (d)). The output power in the central lobe reached 80% from the total lasing output, and its full divergence 2V along axis Y made $2V=\lambda/(N*d)$. Thus, in this range of values P_{in} and L the in-phase supermode lasing was observed. Output power reached 180 mW.

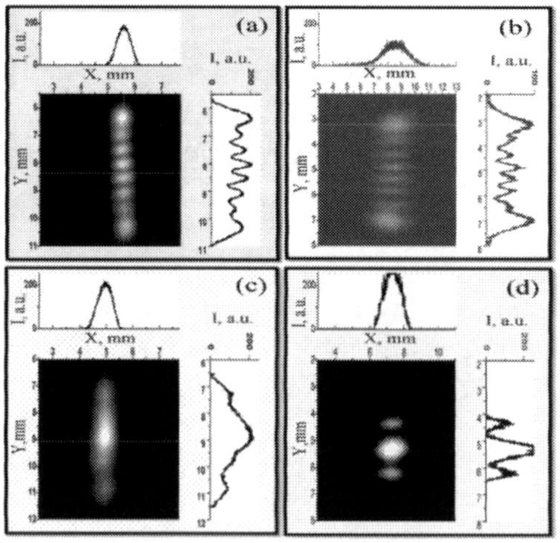

Fig. 1 The radiation intensity transverse distribution: (a) and (c) inside the AD in the condition of collective mode lasing and the super-mode, respectively; (b) and (d) in the far-field zone in the mode of collective mode lasing and in-phase super-mode, respectively.

For an explanation of existence of a supermode lasing the study of the lens thermoinduced in the AD has been done. It has been shown that the supermode lasing may be realized when the laser cavity corresponded to parameters of a degenerate resonator. Based on this explanation the existence conditions of a supermode lasing have been defined for a wide range of pump power values.

References

[1]D. A. Guryev, D. A.Nikolaev, V. B. Tsvetkov, "Nd:GGG disk laser with multipoint spatially periodic optical pumping", Laser Phys. Lett. **13**, 045003 (2016).

https://doi.org/10.1051/epjconf/202024314003

Investigation of sources of subnanosecond pulses of the visible spectral range for medical systems

I.V. Zhluktova[1], *V.A. Kamynin*[1], *N.R. Arutyunyan*[1,2], *A.S. Pozharov*[1], *A.I. Trikshev*[1], *S.A. Filatova*[1], *E.D. Obraztsova*[1], *V.B. Tsvetkov*[1]

1. General Physics Institute of Russian Academy of Scinces, ul. Vavilova 38, 119991 Moscow, Russia

2. National Research Nuclear University «MEPHi», Kashirskoe hwy. 31, 115409 Moscow, Russia

Currently, fiber laser sources of sub-nanosecond pulses used in many fields of science and technology because of their compactness, efficiency, and manufacturability. And the use of stimulated Raman scattering (SRS) provides the generation outside the spectra of rare-earth elements, making it possible to cover the spectral region from 1 to 2.3 µm. Thus, in some works, sources with SRS generation both in solid-state [1] and in fiber lasers [2] were demonstrated. This paper presents an investigation hybrid mode-locked of an ytterbium-doped fiber laser (master oscillator) and with generating the 1-st Stokes. The generation of yellow radiation at a wavelength of 530-580 nm was also obtained using nonlinear crystals from the frequency doubling of the ytterbium-doped laser.

The optical scheme of the experimental setup is shown in Fig.1. The active medium of the MO was pumped using a semiconductor laser diode that emitted at a wavelength of 975 nm. The cavity of the laser was formed by Yb-doped active fiber with a multicomponent cladding (GTWave) and 200 m single-mode fiber. A pair of polarization controllers were added into the laser cavity to obtain NPE mode-locking. The single-walled carbon nanotubes (SWCNT) were fixed between two optical FC/APC- connectors to realize hybrid mode-locking. The radiation was coupled out from the laser cavity using a polarization beam splitter. As a result, the stable hybrid mode-locking was obtained at a wavelength of 1061 nm. The pulse repetition rate of 1 MHz corresponded to the cavity length. The pulse duration was 200 ps. Then, using the Yb-doped fiber amplifier, the average output power at the output of amplifier more than 1 W was obtained.

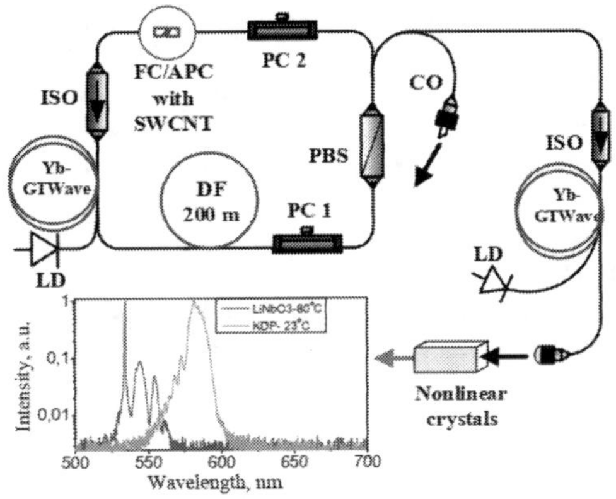

Fig. 1 The optical design of the experimental setup and the spectra obtained by radiation: LD- laser diode, CO- control output, ISO- optical isolator, DF- delay fiber, PC 1,2- polarization controller, PBS- polarized beam splitter.

After amplification of the MO radiation, experiments were performed to the frequency doubling of the ytterbium-doped laser. A KDP(KH_2PO_4), CDA (CsH_2AsO_4), and $LiNbO_3$ crystals were used to get radiation at wavelengths of 530-580 nm. Also, for such crystals as $LiNbO_3$ and CDA, a heating element with a temperature controller was used to achieve 90-degree synchronism. Also, for such crystals as $LiNbO_3$ and CDA, a heating element with a temperature controller was used to attain 90-degree synchronism because the refractive indices of crystals significantly depend on the temperature of samples. The average output power was up to 1 mW. It can be expected that such a laser may find application in medicine to treat red telangiectasia [3] and in ophthalmology [4].

The reported study was funded by RFBR, project number 19-32-90205.

References

[1] M. Frank, S. Smetanin, M. Jelínek, D. Vyhlídal, L. Ivleva, P. Zverev, and V. Kubeček, "Highly efficient picosecond all-solid-state Raman laser at 1179 and 1227 nm on single and combined Raman lines in a BaWO4 crystal, " Optics Lett. **43**, 26715-26721 (2018).

[2] W. Liu, P. Ma, H. Lv. J. Xu, P. Zhou, and Z. Jiang, "General analysis of SRS-limited high-power fiber lasers and design strategy, " Opt. Express., **24**, 26715-26721 (2016).

[3] N.S. Sadick, and R.Weiss, "The Utilization of a New Yellow Light Laser (578 nm) for the Treatment of Class I Red Telangiectasia of the Lower Extremities, " Dermatologic Surg., **28** (1), 21-25 (2002).

[4] K. Inagaki, K. Ohkoshi, S. Ohde, G.A. Deshpande, N. Ebihara, A. Murakami, Jpn. J. Ophthalmol. **59**, 21 (2015).

Passively Q-switched Yb-doped fiber laser based on Ag nanoplates saturable absorber

Pan Wang[1], and Vittorio Scardaci[2,*]
[1]*Department of Precision Instruments, Tsinghua University, Beijing 100084, P. R. China*
Current address: Max Planck Institute for the Science of Light, Staudtstr. 2, 91058, Erlangen, Germany
[2]*Dipartimento di Scienze Chimiche, Università degli Studi di Catania, V.le A. Doria 6, 95125 Catania, Italy;*
E-mail address: vittorio.scardaci@unict.it

Abstract: We experimentally investigated Ag nanoplates as saturable absorber for Q-switched pulse generation in an Yb-doped fiber laser. The pulse train repetition rate increases with the increase of the pump power. At the maximum pump power of 600 mW, the maximum repetition rate and average output power are 184.8 kHz and 10.77 mW, respectively, corresponding to single pulse energy of 58.3 nJ. To the best of our knowledge, it is the first demonstration of the passively Q-switched fiber laser utilizing the material of Ag nanoparticles at the wavelength of 1-μm. Our investigations demonstrate the flexibility of our solution-processed Ag nanoplates-based saturable absorber, making it a promising candidate for a variety of stable and low-cost ultrafast lasers.

Experimental setup and results

Optical deposition procedure is implemented to transfer the Ag nanoplates onto the fiber core area through the thermal effect. The saturable absorber is sandwiched between two fiber connectors, providing simplicity, flexibility and easy integration into the laser oscillator. The modulation depth and saturation incident fluence are measured to be ~5.8% and ~106.36 μJ/cm^2 at 1-μm region, respectively [1].

In the laser cavity, a piece of ~1.8 m moderately ytterbium-doped fiber (Nufern SM-YSF-LO) serves as the gain medium, and is backward pumped by a 980 nm laser diode (LD) through a 980/1060 nm wavelength division multiplexer (WDM). A polarization-insensitive isolator (PI-ISO) is utilized to provide unidirectional operation. Two polarization controllers (PCs) as well as an 8 nm band pass filter centered at 1030 nm, are employed to optimize the Q-switching pulse operation. The rest of the cavity is composed of ~17.7 m standard single mode fibers (SMFs), yielding a total cavity length of ~19.5 m.

Self-started stable Q-switched operation is achieved for a threshold pump power of 180 mW. The repetition rate of the pulse trains ranges from 66.6 to 184.8 kHz when the pump power scales from 210 to 600 mW. The maximum average output power is 10.77 mW, corresponding to the single pulse energy of 58.3 nJ and minimum pulse duration of ~1.01 μs. To the best of our knowledge, it is the first time that the Ag nanoplates saturable absorber are utilized in the 1-μm Yb-doped Q-switched fiber.

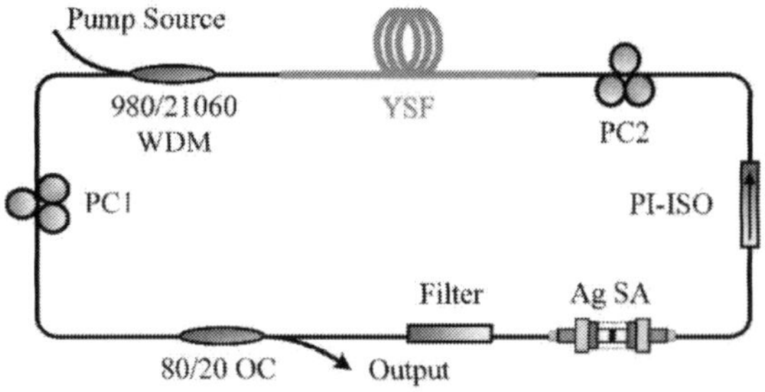

Fig. 1. Scheme of the fiber laser. OC: optical coupler; WDM: wavelength division multiplexer; PC: polarization controller; PI-ISO: polarization-insensitive isolator; Filter: 1030 nm band pass filter; YDF: ytterbium-doped fiber; Ag SA: Ag saturable absorber.

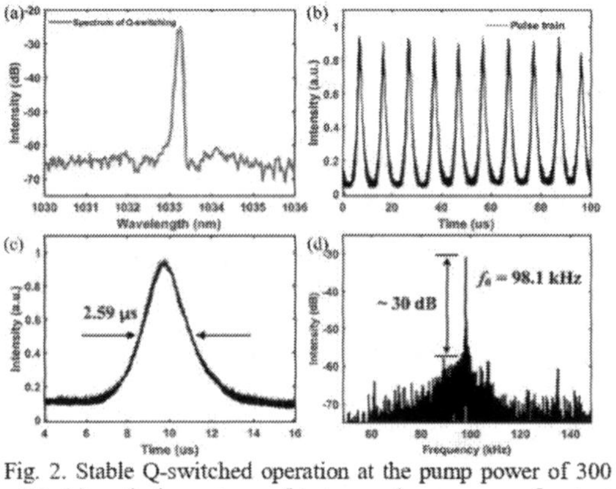

Fig. 2. Stable Q-switched operation at the pump power of 300 mW: (a) optical spectrum; (b) temporal sequences of output pulse recordings; (c) single pulse profile of the Q-switched pulse train; (d) RF spectrum over a 100 kHz span with resolution bandwidth of 10 Hz.

References

[1] Nanophotonics 2020; DOI: 10.1515/nanoph- 2020-0015.

Polarization-domain-wall in a dual-color mode-locked fiber laser

A. Nady[1,2], M. Kemel[1], G. Semaan[1], M. Salhi[1], F. Sanchez[1]

1. Laboratoire de Photonique d'Angers, Université d'Angers, E. A. 4464, 2 Boulevard Lavoisier, 49045 Angers, France

2. Department of Physics, Faculty of Science, Beni-Suef University, 62511 Beni-Suef, Egypt

Generation of rectangular pulses in fiber lasers is of a great interest due to their applications in different fields. Dissipative soliton resonance and polarization domain-wall soliton [1,2] have been introduced as two different mechanisms for rectangular pulse formation.

In this work, we demonstrated domain-wall pulses in a dual-color Er-doped fiber laser. The dual-wavelength operation is realized through the intracavity birefringence induced spectral filtering effect. Stable domain-wall pulses are obtained via the cross-coupling between two lasing wavelengths. This is induced by 20 m high nonlinear fiber (HNLF) incorporated inside a ring cavity consisting of 6 m double-clad fiber Er:Yb and 13 m SMF. In order to study the pulse dynamics, we provide polarization-resolved measurements as well as wavelength-resolved measurements shown in Fig. 1. The optical spectrum shows two spectral lobes centered at ~1547 nm and ~1569 nm, with wavelength separation of ~ 22 nm while the the corresponding temporal trace exhibits a square pulse with an intensity dip, with repetition frequency of ~5.1 MHz corresponding to the cavity round-trip time. The wavelength-resolved measurements (Fig. 1(a)) show that the filtered emissions are two antiphase pulses and the laser emission was found to be interchanging between the two wavelengths; 1547 nm (black trace), and 1569 nm (red trace). This indicates that the total output signal is a combination of two wavelength domain-wall pulses and co-propagate as a non-dispersive unit with a single fundamental repetition rate. On the other hand, the polarization-resolved temporal traces (Fig. 1(b)) of the laser emission display two pulses in the two orthogonal polarization states represented by blue and red traces emerging from the total output signal which is represented by black trace and shows an intensity dip arising from the two orthogonal pulses overlap. Each of the two orthogonal pulses exhibits different pulse width with the same repetition rate. They trap each other through Kerr nonlinearity (cross-phase modulation) and travel as one unit constituting a domain wall.

Polarization-resolved optical spectra in which the signal comes from a single polarization direction are shown in Fig. 1(c) and indicate that each of the polarization components exhibits different spectral distributions and alternately dominates respectively the shorter and longer wavelength after polarization beam splitter. This different wavelength distribution indicates the incoherent coupling between the two polarization components [3]. The presentd results are different from our previous work [4], in which the formation mechanism of the intensity dip is due to the combined effect of gain depletion and chromatic dispersion induced difference in group velocity.

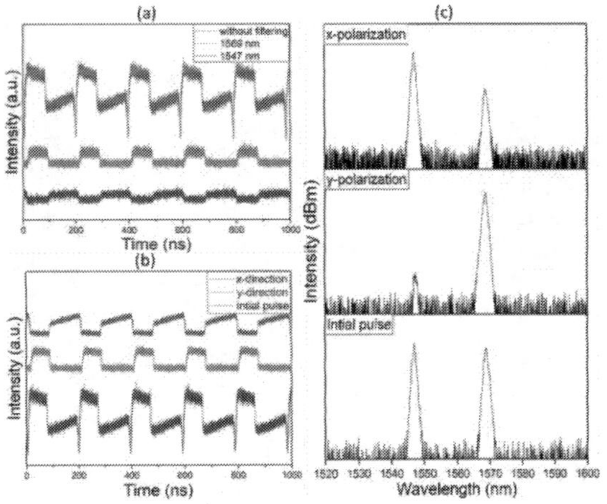

Fig. 1. The Oscilloscope trace of the initial and filtered emission (a), polarization resolved oscilloscope traces (b), and polarization resolved optical spectra (c).

References

[1] W. Chang, A. Ankiewicz, J. M. Soto-Crespo, and N. Akhmediev, "Dissipative soliton resonances," Phys. Rev. A. **78**, 023830 (2008).
[2] P. Wang, K. Zhao, X. Xiao, and C. Yang , "Pulse dynamics of dual-wavelength dissipative soliton resonances and domain wall solitons in a Tm fiber laser with fiber-based Lyot filter," Opt. Express **25**, 30708-30719 (2017).
[3] H. Zhang, D. Y. Tang, L. M. Zhao, and R. J. Knize, "Vector dark domain wall solitons in a fiber ring laser," Opt. Express **18**, 4428-4433 (2010).
[4] Y. Meng, G. Semaan, M. Kemel, M. Salhi, A. Komarov, and F. Sanchez, "Color domains in fiber lasers," Opt. Lett. **43**, 5054-5057 (2018).

Joint spatial profile and frequency conversion of an LP_{07}-fiber mode towards the blue spectral region

Robert Lindberg[1,*], Xiao Liu[2], Siddharth Ramachandran[2], and Valdas Pasiskevicius[1]
1. Department of Applied Physics, Royal Institute of Technology, 10691 Stockholm, Sweden
2. Boston University, 8 St. Mary's St., Boston, USA 02215
Author e-mail address: rl@laserphysics.kth.se

The vast number of light detection and ranging (LIDAR) technologies has led to great advances in both scientific and industrial land-based and airborne applications [1,2]. However, the difficulties in generating high energy pulse in the 440-490 nm region, i.e. the water transmission window [3], has prevented realizations of naval LIDAR systems. Frequency conversion schemes of Yb-fiber [4] and Tm-fiber [5] systems have been proposed to address this issue. They do however respectively suffer from broad output spectra and the need for two free space second harmonic generation (SHG) stages, which could limit the conversion efficiencies.

Our approach instead relies on the power scalability of narrowband four-wave mixing (FWM) in higher order mode fibers to convert well-established commercially available 1064 nm Nd- and Yb-systems to the 970 nm region [6]. The output is then used for SHG in a periodically poled KTP-crystal to reach the water window. Transformation to a Gaussian-like beam profile is simultaneously achieved during the conversion process, which has previously been demonstrated for Gauss-Bessel beams [7], by controlling the phase mismatch [8].

The experimental system revolved around a 650 ps Q-switched Nd:YAG laser at 1064 nm, which was amplified in a double cladding Yb-fiber amplifier, and a tunable Ti:Sa laser at 972 nm to seed the FWM. Both lasers were converted to the LP_{07}-mode, to enable phase-matching, before launched into the multimode fiber. This resulted in a peak power of 5.6 kW at 972 nm, which was focused into a KTP crystal with a poling period of 6.66 μm. Fine-tuning the focusing conditions allowed us to generate remarkably clean on-axis conversion, as shown in Fig. 1 a). The conversion efficiency was however only about 10%, which resulted from that the crystal was placed 2 cm after the focal plane –thus limiting the intensity. SHG closer to the focal plane naturally improved the conversion efficiency, but also led to persistent ring structure.

A numerical investigation of the SHG for the experimental conditions was then undertaken to analyze this trend. The results, shown in Fig. 1 b), revealed that the varying power levels of the pulse envelope generally develop different beam profiles. Acquiring clean on-axis conversion with a non-tophat pulse thus requires all of the power levels to give rise to an averaged profile that is predominantly on-axis, which in our configuration necessitated a displacement away from the focal plane. Future work will therefore be aimed at temporal pulse shaping to enable clean on-axis conversion at higher conversion efficiencies closer to the focal plane.

This work was supported by the US Office of Naval Research under the grants N62909-18-1-2101 and N00014-17-1-2519.

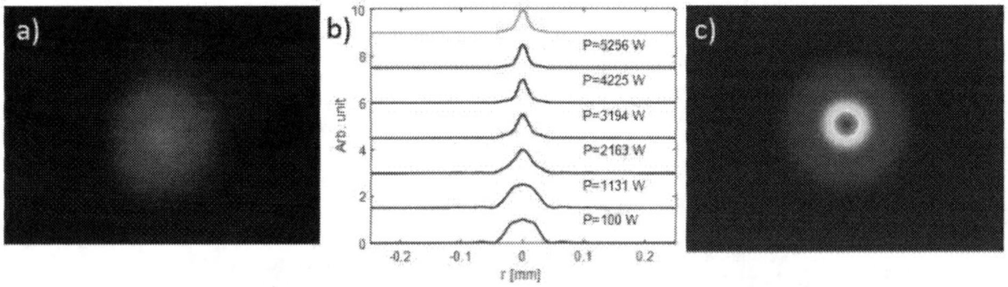

Fig. 1 a) Experimentally obtained on-axis beam. b) Simulated radial profiles at various power levels shown in blue. The green curve shows the resulting integrated profile, whose beam profile is shown in c).

References

[1] K. Omasa et al, "3D lidar imaging for detecting and understanding plant responses and canopy structure," J. Exp. Bot. 58, 881–898 (2007).
[2] R. Rasshofer and K. Gresser, Automotive Radar and Lidar Systems for Next Generation Driver Assistance Functions," Adv. Radio Sci. 3, 205–209 (2005).
[3] N. G. Jerlov, Marine Optics (Elsevier, New York, 1976)
[4] J. Boullet et al, "Millijoule-class Yb-doped pulsed fiber laser operating at 977 nm," Opt. Lett. 35, 1650-1652 (2010).
[5] E. Honea et al, "Pulsed Blue Laser Source Based on Frequency Quadrupling of a Thulium Fiber Laser," Proc. SPIE 9728, 860111 (2013).
[6] J. Demas, et al, "Wavelength-agile high-power sources via four-wave mixing in higher-order fiber modes," Opt. Express 25, 7455-7464 (2017)
[7] A. Piskarskas, et al, "Noncollinear second-harmonic generation in periodically poled KTiOPO4 excited by the Bessel beam," Opt. Lett. 24, 1053-1055 (1999)
[8] R. Lindberg et al, "kW-peak Power Level Blue Light Sources via a Hybrid Fiber and Crystal Nonlinear Process Using Higher Order Modes," Conference on Lasers and Electro-Optics, OSA Technical Digest (2020), paper STh1P.1

https://doi.org/10.1051/epjconf/202024315002

Generation of 60 fs pulses at 780 nm by frequency doubling of Er-doped fiber laser with tunable repetition rate for TPEF imaging

***D. Stachowiak*[1], *J. Bogusławski*[2], *A. Głuszek*[1], *A. Hudzikowski*[1], *Z. Łaszczych*[1], *M. Wojtkowski*[2], *G. Soboń*[1]**

1. Laser & Fiber Electronics Group, Wrocław University of Science and Technology, Wybrzeże Wyspianskiego 27, 50-370 Wrocław, Poland

2. International Centre for Translational Eye Research, Institute of Physical Chemistry, Polish Academy of Sciences, Kasprzaka 44/52, 01-224 Warsaw, Poland

Femtosecond lasers have found an interesting application in the field of biophotonics, in particular in noninvasive diagnostics as optical imaging and multiphoton microscopy, of which example is two-photon excited fluorescent (TPEF) imaging process [1]. Currently, state-of-art TPEF imaging systems utilize commercially-available femtosecond Ti:Sapphire lasers, delivering pulses in the 700-800 nm window, which are quite expensive, complex and large devices [2]. As an alternative one can use a compact Erbium-doped fiber laser which is more simple, robust, maintenance free and cost-effective. By adding a frequency-doubling stage for the second harmonic generation (SHG), femtosecond pulses at 1560 nm wavelength can be converted to ultrashort pulses at 780 nm wavelength, ideally matching two-photon absorption spectra of many fluorophores [3]. Here we demonstrate a frequency-doubled Er-doped fiber laser with tunable repetition rate, optimized for TPEF imaging [4].

The laser system setup is depicted in Fig. 1. The system consists of two stages. The first one is a femtosecond Er-doped fiber laser providing 22 fs pulses with a chosen repetition rate from a range 1÷12 MHz. At the setting of f_{rep} = 1.02 MHz, we have obtained an average power of 6.5 mW corresponding to pulse energy of ~6.37 nJ. Subsequently, those pulses were launched into the second system stage: SHG module. Fig. 2 a) and b) shows the optical second harmonic spectrum and the pulse temporal intensity at f_{rep}=1.02 MHz. Measured central wavelength is 782.6 nm and pulse duration is 60 fs with 1.4 mW of an average power, corresponding to a pulse energy of 1.37 nJ. With increasing the f_{rep} from 1.02 MHz up to 11.9 MHz, we have observed a linear increase of the average power (while preserved pulse energy) and only 2 fs change of the pulse duration.

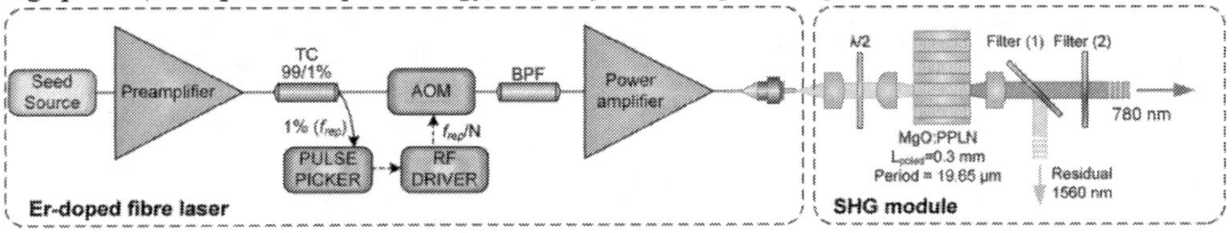

Fig. 1 Scheme of a frequency-doubled fiber laser (TC: tap coupler; AOM: acusto-optic modulator; BPF: bandpass filter).

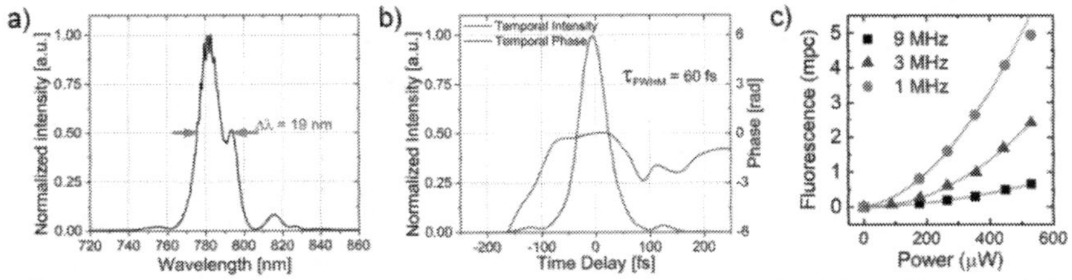

Fig. 2 a) second harmonic spectrum, b) FROG temporal 780 nm pulse intensity (solid red line) and its temporal phase (solid blue line), c) Fluorescence intensity of ex vivo rat skin in mean photon counts per pixel (mpc) as a function of excitation average power for 1, 3, and 9 MHz repetition rates.

Thanks to the built-in pulse-picker, the laser can be used in a wide range of repetition frequencies with preserved pulse energy and duration. It is an important feature allowing for custom setup optimization for specific requirements of the experiment (e.g. lifetime of fluorophores). Additionally, as it can be seen in Fig. 2c), the fluorescence intensity of rat skin sample increases with decreased pulse repetition rate, indicating that it is possible to obtain the same fluorescence intensity with 1 MHz excitation using ~3 times less power compared with 9 MHz excitation, thus reducing the thermal damage probability. The system was successfully applied for TPEF imaging of various biological samples confirming the benefit from adjustable pulse repetition rate function. This work was financed by Foundation for Polish Science (First TEAM/2017-4/39, TEAM TECH/2016-3/20, MAB/2019/12).

References

[1] W. Denk, J. H Strickler, W. W. Webb, "Two-photon laser scanning fluorescence microscopy," Science 248(4951), 73-76 (1990).
[2] G. Palczewska, et al., "Two-photon imaging of the mammalian retina with ultrafast pulsing laser," JCI Insight 3, e121555 (2018).
[3] S. Boivinet, P. Morin, J. P. Yehouessi, S. Vidal, G. Machinet, J. Boullet, "3.5nJ femtosecond pulses at 792 nm generated by frequency doubling of an all-PM fiber high energy 48 fs laser," Proc. SPIE 10897, Fiber Lasers XVI: Technology and Systems, 108971J (2019).
[4] D. Stachowiak, J. Bogusławski, A. Głuszek, Z. Łaszczych, M. Wojtkowski, G. Soboń, "Frequency-doubled femtosecond Er-doped fiber laser for two-photon excited fluorescence imaging", arXiv:2005.00927 [physics.optics] (2020).

 https://doi.org/10.1051/epjconf/202024316001

Mid-infrared electric field sampling approaching single-photon sensitivity

C. Hofer[1,2], D. Gerz[1,2], M. Högner[1], T. P. Butler[3], C. Gaida[4], T. Heuermann[5,6], M. Gebhardt[5,6], N. Karpowicz[8], J. Limpert[5,6,7], F. Krausz[1,2], I. Pupeza[1,2]

1. Max Planck Institute of Quantum Optics, Hans-Kopfermann-Str. 1, D-85748 Garching, Germany

2. Ludwig Maximilians University Munich, Faculty of Physics, Am Coulombwall 1, D-85748 Garching, Germany

3. University of Maryland, Institute for Research in Electronics and Applied Physics, College Park, MD, United States

4. Active Fiber Systems GmbH, Ernst-Ruska Ring 11, D-07745 Jena, Germany

5. Institute of Applied Physics, Abbe Centre of Photonics, Friedrich-Schiller Univ. Jena, Albert-Einstein Str. 15, D-07745 Jena, Germany

6. Helmholtz-Institute Jena, Frobelstieg 3, D-07743 Jena, Germany

7. CNR NANOTEC Institute of Nanotechnology, IT-73100 Lecce, Italy

8. Fraunhofer Institute for Applied Optics and Precision Engineering, Albert-Einstein Str. 7, D-07745 Jena, Germany

Electro-optic sampling (EOS) [1,2] enables measurements of light fields with sub-cycle temporal resolution. Here, we show the feasibility of mid-infrared (MIR), field-resolved detection with nearly single-photon sensitivity. This is facilitated by the thorough optimization of our high-power, Thulium (Tm)-based laser system and the EOS detection [3]. The experimental setup is shown in Fig. 1a and is described in detail in Ref. [4].

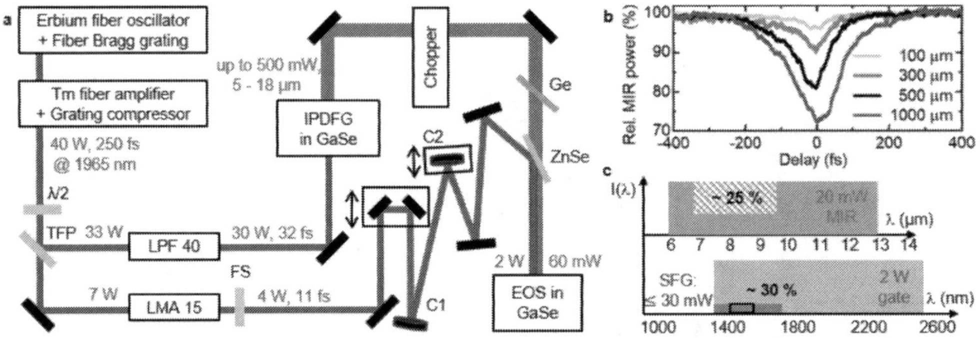

Fig. 1 (a) Experimental setup. TFP: Thin film polarizer, LPF: Large pitch fiber, LMA: Large mode area, C1, C2: concave mirrors. (b) Results of MIR depletion measurements for several EOS crystal thicknesses. (c) Sketch of the overall MIR photon detection efficiency.

The output from the Tm-fiber amplifier is further compressed with photonic crystal fibers in two channels. 30 W, 32-fs-pulses are used to drive intra-pulse difference frequency generation (IPDFG) in Gallium Selenide (GaSe), generating high-power, few-cycle MIR transients. A mechanical chopper modulates the beam at 9 kHz to allow for lock-in detection of the EOS traces. We scan the delay of the 2 W, 11 fs EOS gate pulses with a mechanical stage. A telescope, consisting of two concave mirrors, is used to optimize the spatial overlap of the MIR fields and gate pulses in the EOS crystal. They are collinearly combined for EOS detection with an uncoated Zinc Selenide (ZnSe) plate in Brewster´s angle.

We characterize the detection efficiency by measuring the delay-dependent MIR depletion (Fig. 1b). The maximum, spectrally integrated up-conversion efficiency is 25 %, for a 1 mm thick GaSe EOS crystal. A spectral filter is employed for optimization of the detection dynamic range [5], allowing – together with reflection losses – for the detection of ~ 30 % of all sum-frequency photons. As indicated in Fig. 1c, the spectrally-resolved depletion surpasses 50 %. Therefore, for those wavelengths with largest depletion, and with balanced detection close to the near-infrared shot noise limit, MIR photon numbers as low as 10/second can be detected in a field-resolved manner.

Together with record MIR-powers (> 10^{18} photons/second at the detector), this paves the way to linear, field-resolved measurements over up to 9 (field) and 18 (intensity) orders of magnitude. This unprecedented sensitivity and dynamic range allow us to push field-resolved spectroscopy [6] in the MIR to its fundamental limits.

References

[1] Q. Wu and X.-C. Zhang, "Free-space electro-optic sampling of terahertz beams," Appl. Phys. Lett. **67**, 3523 (1995).
[2] G. Gallot and D. Grischkowsky, "Electro-optic detection of terahertz radiation," J. Opt. Soc. Am. B **16**, 1204 (1999).
[3] C. Hofer et al., "Quantum-Efficiency and Bandwidth Optimized Electro-Optic Sampling," CLEO Europe (2019), DOI: 10.1109/CLEOE-EQEC.2019.8873083.
[4] T. P. Butler et al., "Watt-scale 50-MHz source of single-cycle waveform-stable pulses in the molecular fingerprint region," Opt. Lett. **44**, 1730 (2019).
[5] M. Porer et al., "Shot noise reduced terahertz detection via spectrally postfiltered electro-optic sampling," Opt. Lett. **39**, 2435 (2014).
[6] I. Pupeza et al., "Field-resolved infrared spectroscopy of biological systems," Nature **577**, 52 (2020).

EPJ Web of Conferences **243**, 16002 (2020) https://doi.org/10.1051/epjconf/202024316002
EUROPHOTON 2020

Mid-infrared waveform measurement by rapid mechanical scanning

Philip Jacob[1], Alexander Weigel[1,2], David Gröters[2], Theresa Buberl[1], Michael Trubetskov[1], Marinus Huber[1,2], Joachim Heberle[3], Ioachim Pupeza[1,2]

1. Max-Planck-Institut für Quantenoptik, Hans-Kopfermann-Str. 1, Garching 85748, Germany
2. Fakultät für Physik, Ludwig-Maximilians-Universität München, Am Coulombwall 1, Garching 85748, Germany
3. Department of Physics, Experimental Molecular Biophysics, Freie Universität Berlin, Arnimallee 14, Berlin 14195, Germany

Many applications of ultrafast lasers, such as pump-probe spectroscopy [1], 2D femtosecond spectroscopy [2] and field resolving electro-optic sampling (EOS) [3] demand the scan of one or more optical delays between laser pulses. Conventionally, this is realized by moving a retroreflector with a mechanical stage, typically restricting the scan rate to the Hz regime at best. Laser intensity noise, however, has major contributions from sub-kHz components that directly influence the scan results. Performing full optical scans at multi-kHz rates can eliminate signal distortion by such low-frequency noise components. In addition, repeating full optical scans on sub-ms time scales allows to follow chemical reactions in real time [4]. Several approaches have been pursued in the past to achieve picosecond-delay optical scans at kHz rates, including for example electronically controlled synchronization of two laser oscillators [5] and acousto-optic delay generation [6]. Many of the proposed techniques are either costly and complex or place restrictions on the wavelength and bandwidth of the laser pulses.

In this work we amplify the motion of a resonantly oscillating piezo with a sonotrode to perform forward/backward scans of a mirror over a 1.5-ps range at an effective rate of 38 kHz. The sonotrode scanner provides a simple and cost-effective way to upgrade a conventional femtosecond experiment to ultrafast scan speeds without wavelength or spectral bandwidth limitations. Its application has been previously shown for time-resolved FTIR spectroscopy with a non-coherent light source [4]. Here we demonstrate for the first time the use of a sonotrode in a complex non-linear optical experiment with femtosecond pulses: The mapping of the electric field of a few-cycle mid-infrared (mid-IR) pulse generated by intra-pulse difference frequency generation of compressed pulses from an Yb:YAG oscillator (28 MHz, 60 W, 17 fs) by EOS [3]. We use the interferometric modulation of an auxiliary continuous-wave beam to precisely track the resonant motion of the moving mirror in real time (Fig. 1a). By applying a predictor-corrector algorithm we were able to perform consistent calibration of the delay-time axis over several seconds of integration time, with a time-precision reaching 111 attoseconds for a single scan, and 14 attoseconds for 100 scans. Accurate knowledge of the delay axis is necessary for harnessing the advantages of fast scanning by averaging multiple scans or tracking signal evolution from chemical reactions in a scan series. The inset in Fig. 1b shows how oscillations which are overpowered by noise in a single scan become visible upon averaging. The baseline noise scales with the square root of the number of traces averaged (Fig. 1c), so that with 1 s acquisition time, a dynamic range of >10^8 is reached.

In conclusion, we have shown EOS of few-cycle mid-IR pulses with a simple mechanical multi-kHz scanner, while maintaining attosecond precision and reproducibility for the delay axis.

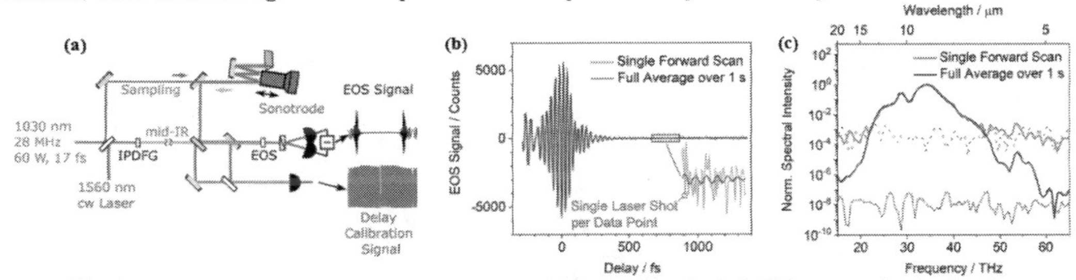

Fig. 1. (a) Schematic diagram of the experimental setup. (b) Comparison of a single EOS scan mapping the electric field of the mid-infrared pulse with the mean of all traces acquired over 1s (i.e., 38000 scans). (c) Mid-IR intensity spectra obtained by Fourier transformation of the traces in (b) (Dashed lines represent baseline noise).

References

[1] A. Zewail, "Femtochemistry: Atomic-Scale Dynamics of the Chemical Bond Using Ultrafast Lasers (Nobel Lecture)", Angew. Chem. Int. Ed. **39**, 2586-2631 (2000).
[2] D.M. Jonas, "Two-dimensional femtosecond spectroscopy", Annu. Rev. Phys. Chem. **54**, 425-463 (2003).
[3] I. Pupeza *et al.*, "Field-resolved infrared spectroscopy of biological systems," Nature **577**, 52–59 (2020).
[4] B. Süss, F. Ringleb, and J. Heberle,."New ultrarapid-scanning interferometer for FT-IR spectroscopy with microsecondtime-resolution," Rev. Sci. Instrum. **87**, 063113 (2016)
[5] S. Kray, F. Spöler, T. Hellerer, and H. Kurz, "Electronically controlled coherent linear optical sampling for optical coherence tomography," Opt. Express **18**, 9976-9990 (2010)
[6] O. Schubert, M. Eisele, V. Crozatier, N. Forget, D. Kaplan, and R. Huber, "Rapid-scan acousto-optical delay line with 34 kHz scan rate and 15 as precision," Opt. Lett. **38**, 2907-2910 (2013)

Time-Locked Multi-Color Single-Aperture Fiber Sources via Soliton Self-Mode Conversion

Havva Begüm Kabagöz, Aku Antikainen and Siddharth Ramachandran
Boston University, 8 St. Mary's Street, Boston, MA, 02215, USA

Multi-color time-synchronized sources are useful for applications such as Raman spectroscopy [1], and mid-IR generation [2]. Multi-color ultrashort pulse sources from fibers, however, suffer from relative timing-jitter because different colors typically travel at different group velocities, hence experiencing differing levels of RIN to delay variations. Here, we show that the natural intermodal group-velocity locking characteristic of soliton self-mode conversion (SSMC), a process recently discovered in multimode fibers [3, 4], can help generate temporally locked energetic ultrashort pulses across a range of wavelengths, spectrally separated, typically, by one or more Raman Stokes shifts. Control of the wavelength at which the emissions occur, and the number of pulses that can be generated, depends primarily on the group-index separation of the modes in a step-index fiber, which can be controlled via fiber core diameter.

Our pump laser is at 1045-nm with 100-fs pulse width and 1-MHz repetition rate (KMLabs YFi). Spatially sculpting this pump into the $LP_{0,20}$ mode of a 97-μm core diameter, 0.34-NA step-index fiber (Fig. 1.a,b) with an SLM, results in two well-defined ~75-fs pulses in the $LP_{0,20}$ and $LP_{0,19}$ modes, respectively, as they are group velocity matched across one Raman Stokes shift (Fig. 1.c,d). Frequency-doubling the pulses and studying their cross-correlations reveals a jitter of only 13 fs between these two 75-fs pulses, demonstrating temporal locking. The versatility of this technique is illustrated with an identical fiber but now with 70-μm core diameter (Fig. 1.e). When the pump is in the $LP_{0,17}$ mode (Fig. 1.f), three modes are group-index matched across two Raman Stokes shifts (Fig. 1.g) yielding a three-color source (Fig. 1.h) [5]. And, when the pump mode is in the $LP_{0,14}$ mode instead, four-color emission is obtained (Fig.1.i,j,k).

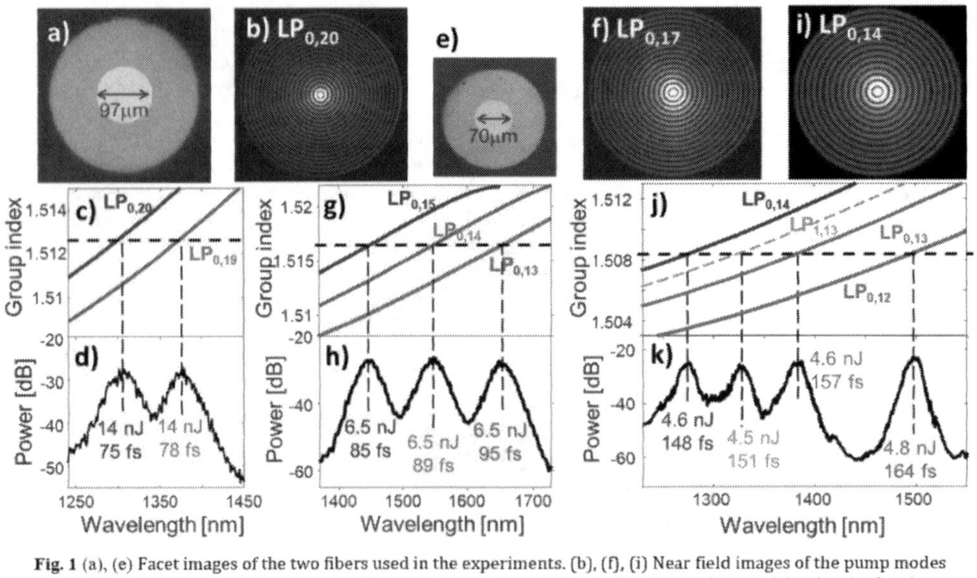

Fig. 1 (a), (e) Facet images of the two fibers used in the experiments. (b), (f), (i) Near field images of the pump modes used in the three experiments. (c), (g), (j) Group-index vs. wavelength for the modes participating in the three respective experiments. (d), (h), (k) the corresponding spectra of the multi-color emissions of the three experiments with pulse energies and autocorrelator-measured pulse widths shown for each peak. Vertical dashed lines show the excellent match between the experimentally observed spectral positions of the pulses, and the corresponding wavelengths at which the modes are group-velocity matched.

Group-velocity matching *across* spatial modes of a fiber helps obtain multiple ~100-fs pulses with up to 14-nJ pulse energies and multiple colors spanning > 200 nm emitted from a single (fiber) aperture. This is enabled by the process of soliton self-mode conversion, which leverages the new degrees of freedom afforded by multimode nonlinear optics to develop novel ultrashort pulse fiber sources.

[1] M.D. Duncan et al., "Scanning coherent anti-Stokes Raman microscope." Opt. Lett. **7**, 350 (1982).
[2] D.G. Winters et al., "Sub-ps fiber-based soliton-tuned mid-IR source in the 9.7–14.9μm wavelength..." Opt. Lett. **35**, 2179 (2010).
[3] L. Rishøj et al., "Soliton self-mode conversion: revisiting Raman scattering of ultrashort pulses," Optica **6**, 304 (2019).
[4] L. Rishøj et al., "Jitter-free, dual-wavelength, ultrashort-pulse, energetic fiber sources using SSMC," Opt. Express **28**, 4333 (2020)
[5] H. B. Kabagöz et al., "Time-Synchronized 3-Color Single-Aperture Fiber Sources via SSMC," CLEO, p. SM1P.7 (2020).

Demonstration of supercontinuum and frequency shifted solitons pumped at 1.56 μm as seed sources for Tm-doped fiber amplifiers

***Olga Szewczyk*[1], *Aleksander Głuszek*[1], *Tadeusz Martynkien*[2]**
***Karol Tarnowski*[2], *Paweł Mergo*[3], *Grzegorz Soboń*[1]**

1. *Laser & Fiber Electronics Group, Wrocław University of Science and Technology, 50-370 Wrocław, Poland*
2. *Department of Optics and Photonics, Wrocław University of Science and Technology, 50-370 Wroclaw, Poland*
3. *Laboratory of Optical Fiber Technology, Maria Curie-Skłodowska University, 20-031 Lublin, Poland*

Stable, low noise and coherent light sources operating in near- and mid-infrared regime draw much attention recently due to their potential scientific, medical and industry applications. Nonlinear effects that occur in microstructured silica fibers, e.g. soliton self-frequency shift (SSFS) [1] or supercontinuum (SC) generation [2] can be exploited to build such sources as an alternative to lasers based on conventional gain media, e.g. Ho- or Tm-doped fibers combined with different mode-locking techniques. Nowadays, the possibility of using dispersion-engineered fibers with their structure consisting of numerous air-holes in the cladding gives the opportunity to control the nonlinearities in such fibers which makes it a powerful tool for nonlinear optics. Since both SSFS and SC were already used as seed sources for Tm-doped amplifiers [1,3] (as an alternative to a Tm-doped mode-locked laser), the aim of the presented work is to directly compare their performance in terms of power, noise, coherence and stability.

In our experiment (Fig. 1(a)), an Er-doped fiber laser operating at 1.56 μm, emitting 50 fs pulses at 125 MHz repetition rate was used as a pump source. The laser beam was coupled into two nonlinear fibers – an all-normal dispersion (ANDi) fiber generating SC and anomalous dispersion fiber fiber that generated tunable solitons in the range up to 2 μm. The noise dynamics of the pulses emitted at the outputs of both fibers were then analyzed by measuring degree of coherence, shot-to-shot noise (by means of dispersive Fourier transform) and relative intensity noise (the last presented in Fig. 1(b)). The obtained results indicate high coherence level as well as good noise properties for both nonlinear fibers, yet with slightly better performance for the SSFS. Both types of sources were used as seeds for an all-fiber Tm-doped fiber amplifier, leading to 88 fs pulses with average power of 365 mW for the SSFS and 92 fs pulses with average power of 360 mW for the SC. What is more, the emission wavelength of the soliton could have been adjusted and the pulses of a duration 80 fs and optical power 400 mW could have been achieved. Adding a low shot-to-shot noise confirmed with DFT, high coherence and low RIN, we show that Raman-shifted solitons are excellent seed sources for Tm-doped fiber amplifiers, outperforming ANDi-SC. This work was financed by Foundation for Polish Science (First TEAM/2017-4/39).

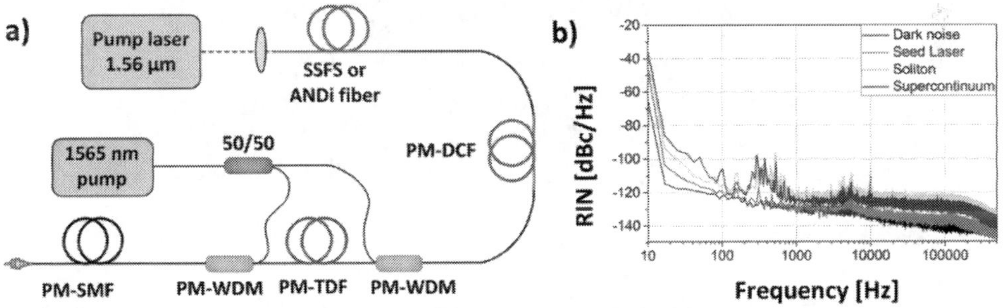

Fig. 1 Schematic of the experimental setup (ANDi – all-normal dispersion fiber, SSFS – soliton self-frequency shift fiber, PM-DCF – polarization maintaining dispersion compensating fiber, PM-WDM – PM wavelength division multiplexer, PM-TDF – PM thulium-doped fiber, PM-SMF – PM single-mode fiber (a). Relative intensity noise of the soliton and supercontinuum and the seed laser comparison (b).

References

[1] G. Soboń, T. Martynkien, D. Tomaszewska, K. Tarnowski, P. Mergo, and J. Sotor, "All-in-fiber amplification and compression of coherent frequency-shifted solitons tunable in the 1800–2000 nm range," Photon. Res. **6**, 368-372 (2018).
[2] K. Tarnowski, T. Martynkien, P. Mergo, J. Sotor, and G. Soboń, "Compact all-fiber source of coherent linearly polarized octave-spanning supercontinuum based on normal dispersion silica fiber", Sci. Rep. **9**, 12313 (2019).
[3] A. Rampur, Y. Stepanenko, G. Stępniewski, T. Kardaś, D. Dobrakowski, D.-M. Spangenberg, T. Feurer, A. Heidt, and M. Klimczak, "Ultra low-noise coherent supercontinuum amplification and compression below 100 fs in an all-fiber polarization-maintaining thulium fiber amplifier," Opt. Express **27**, 35041-35051 (2019).

Observation of supercontinuum spiral emission in optical fibers

F. Mangini[1], M. Ferraro[2], M. Zitelli[2], A. Niang[1], A. Tonello[3], V. Couderc[3], S. Wabnitz[2]

1. *Department of Information Engineering (DII), University of Brescia, Via Branze 38, 25123 Brescia, Italy.*
2. *Department of Information Engineering, Electronics and Telecommunications (DIET), Sapienza University of Rome, Via Eudossiana 18, 00184 Rome, Italy.*
3. *Université de Limoges, XLIM, UMR CNRS 7252, 123 Avenue A. Thomas, 87060 Limoges, France*

Research on multimode (MM) optical fibers has attracted an increasing interest in recent years, thanks to their application potential [1]. Studies of nonlinear propagation MM fibers are also interesting for a fundamental viewpoint. Thanks to their unique properties, unexpected and yet to be fully understood wave phenomena have been observed in MM fibers. A remarkable example is the appearance of energetic and multi-octave supercontinuum spectra, extending from the visible to the mid-IR, at sufficiently high pumping intensities [2]. It has been shown that pump pulses with 100 ps temporal durations and peak powers around 200 kW lead to supercontinuum generation in MM fibers with a graded-index (GRIN) core profile [3]. In this work, we reveal that when using femtosecond pump pulses, one has access to a different regime, where the peak power inside the MM fiber may reach and even surpass the critical power P_{cr} for catastrophic self-focusing (this is about 10 MW in fused-silica). When these peak power levels are reached, the propagating light beam undergoes a so-called filamentation phenomenon, leading to the so-called conical emission (CE) [4]. In our experiments, we report for the first time the formation of a supercontinuum spiral emission (SSE), where the spectral components of the output beam are spatially separated in the far-field emitted at the fiber output. We used both a commercial 50/125 GRIN fiber and a 105/125 step-index fiber. Such an unexpected far-field beam shape was obtained by raising the input peak power above the critical threshold P_{cr}. Our pump laser delivered 120 fs pulses at 1030 nm with up to 25 MW peak power. The repetition rate was set to 2 kHz, and the beam was focused by a 19 mm lens, producing an 8 μm beam waist at the fiber input facet. We show here results with a 2 cm long multimode standard GRIN and step-index fiber, relative index difference $\Delta = 0.0103$ (cladding refractive index is 1.457), and chromatic dispersion $\beta_2 = -22$ ps^2/km.

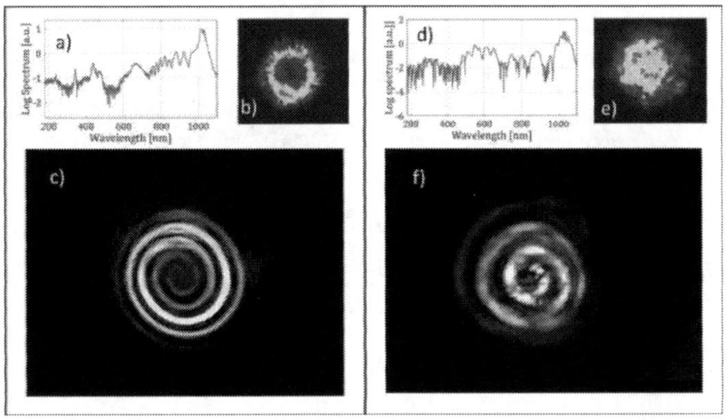

Fig. 1 Experimental results obtained in a 50/125 GRIN fiber (left panel) and Step-index fiber (right panel); a) Output spectrum at 25 MW input peak power (logarithmic scale); b) near field; c) far field.

In Figs.1.a, the output spectrum shows an ultrabroadband distribution, ranging from the UV up to the pump wavelength. The near-field profile at the fiber output shows that the supercontinuum only propagates inside the cladding (see Figs. 1.b). Finally, the far-field SSE is shown in Figs. 1.c. In Fig1 d)-f) we show how the spectrum, the near- and far-field profiles change when the input coupling condition is slightly changed. The spiral-shaped emission is maintained. However, the supercontinuum distribution no longer shows a spatially separated spectral distribution. Moreover, the near-field manifests the typical multimodal speckled profile. A theoretical model able to describe SSE generation is being developed.

This work was supported by the European Research Council (grant No. 740355).

References

[1] K. Krupa, et al., "Multimode nonlinear fiber optics, a spatiotemporal avenue," APL Photonics **11**, 110901 (2019).
[2] L. G. Wright, et al., "Ultrabroadband Dispersive Radiation by Spatiotemporal Oscillation of Multimode Waves" PRL 115, 223902 (2015).
[3] G. Lopez-Galmiche, et al., " Visible supercontinuum generation in a graded index multimode fiber pumped at 1064 nm" Opt. Lett. 41, 11 (2016).
[4] E. O. Smetanina, et al., "Supercontinuum conical emission accompanying filamentation of a femtosecond laser pulse in fused quartz" J. Opt. Technol. 77, 7 (2010).

 https://doi.org/10.1051/epjconf/202024317004

Polarization-dependent relative intensity noise of fiber supercontinuum sources

Dirk-Mathys Spangenberg, Benoît Sierro, and Alexander M. Heidt
Institute of Applied Physics, University of Bern, Sidlerstrasse 5, 3012 Bern, Switzerland

Highly coherent and low noise supercontinuum (SC) sources based on nonlinear spectral broadening of femtosecond pulses in all-normal dispersion (ANDi) fibers are attractive for many applications in ultrafast photonics, such as nonlinear bio-photonic imaging, ultrafast spectroscopy, coherent X-ray generation, and low-noise ultrafast fiber laser development, amongst others [1]. The noise properties of the SC source are of particular importance as fluctuations translate to intensity noise, pulse duration noise, or timing jitter, affecting sensitivity, resolution, or synchronisation of ultrafast experiments. While ANDi SC exhibit superior noise properties compared to their conventional counterparts pumped in the anomalous dispersion regime, recent theoretical studies suggest that polarization modulation instability (PMI) can severely degrade their stability [2].

Here we present, for the first time, detailed experimental polarization-dependent relative intensity noise (RIN) measurements of ANDi SC sources based on both polarization-maintaining (PM) and non-PM fibers, and compare them to a conventional SC source (Fig. 1). All fibers were pumped with an ultrafast Er:fiber laser (80 fs, 40 MHz, 0.05% RIN), generating SC with comparable spectral bandwidths in the range 1.2 – 2.2 µm. A rotating half-wave plate in front of the fiber and a synchronized analyzer at the fiber exit controlled the plane of pump pulse and detection polarization with respect the fiber's principal axes. Integrated RIN values from 0-18 MHz were measured using a photodiode and electronic spectrum analyzer with an angular resolution of approximately 0.2°.

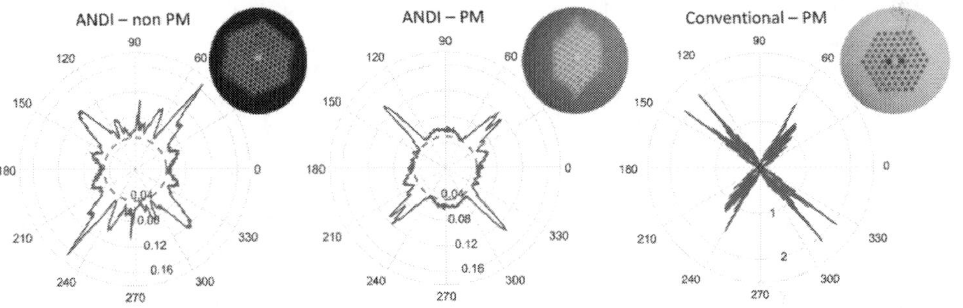

Fig. 1: Polarization-dependent RIN measurements of SC generated in ANDI PM and non-PM fiber as well as in a commercial conventional nonlinear PM fiber pumped in the anomalous dispersion regime (NKT PM-1550-01). The polar plot shows the polarization angle of the pump pulses with respect to the principal fiber axes, the RIN values are integrated from 0-18 MHz and given in percent. The red-dotted line shows the RIN of the pump laser.

Figure 1 presents polarization-dependent RIN measurements for each fiber sample. While the SC generated in the non-PM ANDi fiber shows complex, polarization dependent RIN features that are sensitive to environmental conditions, amplifying the pump laser noise by a factor of up to 3, the PM design completely suppresses noise amplification by PMI around the fiber's principal axes with high environmental stability. In this case, the RIN of the SC shows only very weak dependence on pump pulse polarization and is virtually identical to the RIN of the pump laser, except for sharp peaks at 45° to the principal axis where the noise of the pump laser is amplified by a factor of 2. In contrast, the RIN of the conventional SC source rises sharply for off-axis polarization reaching maximum RIN values of 2 %, which corresponds to a noise amplification factor of about 40.

Our results demonstrate that the ANDi PM fiber design can effectively suppress nonlinear noise amplification by PMI during SC generation. In addition, ANDi-PM SC sources prove to be robust against input polarization fluctuations, while the noise of conventional SC sources, even when based on PM-fiber, rises sharply for slight misalignment between pump laser polarization and fiber axes. In combination with the previously demonstrated minimization of both quantum and technical noise amplification in ANDi fibers [3,4], our results underline the substantial potential of the ANDi fiber design for applications that require ultra-low noise, high quality SC sources.

References

[1] A. M. Heidt, A. Hartung, and H. Bartelt in, *The Supercontinuum Laser Source: The Ultimate White Light*, R. Alfano ed. (Springer, New York, 2016), Chapter 6, pp. 247–280.

[2] I. Bravo Gonzalo, R. D. Engelsholm, M. P. Sørensen, and O. Bang, "Polarization noise places severe constraints on coherence of all-normal dispersion femtosecond supercontinuum generation," Sci. Rep. **8**, 6579 (2018).

[3] A. M. Heidt, J. S. Feehan, J. H. V. Price, and T. Feurer, "Limits of coherent supercontinuum generation in normal dispersion fibers," J. Opt. Soc. Am. B **34**, 764–775 (2017).

[4] B. Sierro and A. M. Heidt, "Noise amplification in all-normal dispersion fiber supercontinuum generation and its impact in ultrafast photonics applications," OSA Continuum (in press).

Dependence of Raman scattering on the orbital angular momentum of light

Xiao Liu, Aku Antikainen, and Siddharth Ramachandran
Boston University, 8 Saint Mary's St., Boston, Massachusetts, 02215, USA
Email: xlewis@bu.edu

Raman scattering has found a variety of applications in areas such as amplification and spectroscopy [1,2], but it is also sometimes undesired, for instance, in power-scaling fiber lasers or in generating low-noise fiber-based quantum sources [3]. Controlling Raman scattering in fibers is thus important, but also a topic intensively studied, since its properties are generally well known. For instance, Raman gain is proportional to the nonlinear overlap integral $\int |E_P|^2 |E_s|^2 dA$, where E_P and E_s are the electric fields for pump and Stokes light, respectively [1] – a quantity that can be influenced by mode effective area. Also well-known is the fact that Raman gain is highest for co-polarized light [1]. Here, we show that orbital angular momentum (OAM) carried by light, which manifests as a helical phase $e^{iL\phi}$ of its field, where L is the topological charge, can also control Raman scattering. This seems counter-intuitive since the nonlinear overlap integral above indicates that Raman depends on intensity instead of electric field, so phase should not play a role.

We consider two pump cases – circularly and linearly polarized OAM beams. While the circularly polarized OAM beam behaves like any other fiber mode, a linearly polarized OAM beam is a superposition of left and right-handed circularly polarized $\sigma^{\pm}$ fiber modes with the same L. Since these two modes are non-degenerate in effective index [4], we obtain a state that experiences circular birefringence Δn_{eff}, and hence optical activity (OA). Upon fiber-propagation, the mode remains linearly polarized, but rotates in polarization orientation [5].

In our experiment, the pump laser is a Q-switched Nd:YAG laser amplified with a 1.5-m long Yb-doped fiber. SLMs and polarization optics are used to sculpt and couple light into the desired OAM state of a ring-core fiber ($\Delta n = 0.035$, ring inner diameter = 21 μm, ring thickness = 5 μm) designed to stably propagate such modes [6]. Figure 1. (a) shows the measured near-field images for 8 different OAM modes – note that they have similar shapes and sizes, indicating that their nonlinear overlap integral for Raman scattering is similar. Figure 1. (b) shows spectra for two pumping cases of same input peak power (10 kW) – when the pump is in a circularly polarized OAM state ($L = 22, \sigma^+$) and when it is in the OA state ($L = 22, \hat{x}$). At one Stokes shift away from the pump (~1115 nm), it is clear to see that the Raman power of the OA state is 15 dB lower than that of the circularly polarized OAM input [7]. Spectral features close to the pump are four wave mixing peaks not described here – that they are comparable in both cases just confirms that the differing Raman strengths mentioned above are not artifacts from experimental variations. To study the origins of differing Raman gain responses just by changing polarization from circular to linear, we simulate this effect [8]. Figure 1. (c) shows that Raman gain for the OA state is maximized for a linearly polarized $L = 21$ mode – i.e. also an OA state but exactly one mode order lower than the pump. More generally, Raman gain appears to depend on mode order, which seems counter to Fig. 1. (a) that suggests Raman gain should be similar for all modes. This is resolved by considering the inherent polarization rotation of OA states with wavelength-dependent beat lengths, as shown in Fig. 1. (d). Polarization walk-off strongly diminishes Raman gain for a mode of the same L, and instead maximizes co-polarized interactions when the beat-lengths of the pump and Stokes modes match.

This demonstrates a dependence of Raman gain on the inherent chirality of optical fiber modes, and hence potentially provides an alternative means of tailoring Raman scattering in fibers.

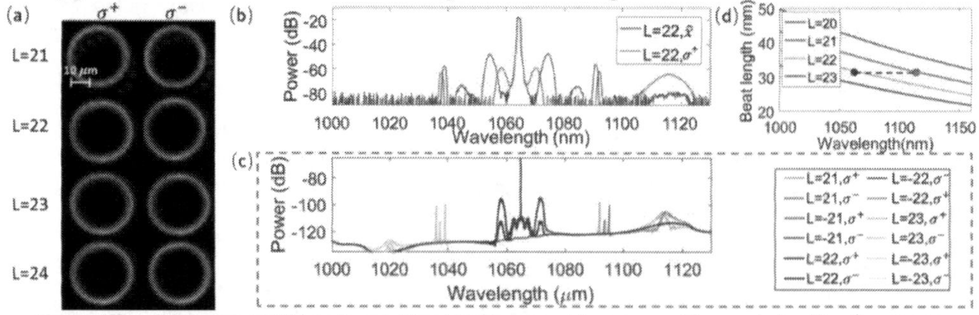

Fig. 1 (a) Near-field images of OAM modes considered (b) Experimental spectrum with pump in $L = 22$; σ^+ *or* $\hat{x}$ state (c) Simulated spectra for pump in $L = 22$; $\hat{x}$ (i.e. OA) state (d) OA polarization evolution beat lengths for L=20-23

References

[1] G. P. Agrawal, *Nonlinear Fiber Optics*, 3rd. ed., (Academic Press, Boston, 2001).
[2] A. Zumbusch et al., "Three-dimensional vibrational imaging by coherent anti-Stokes Raman scattering," *Phys. Rev. Lett.* **82** 4142 (1999)
[3] X. Li et al., "Optical fiber-source of polarization-entangled photons in the 1550 nm telecom band," *Phys. Rev. Lett.* **94** 053601 (2005).
[4] S. Ramachandran et al., "On the scalability of ring fiber designs for OAM multiplexing," *Opt. Express* **23**, 3721 (2015)
[5] A. P. Greenberg et al., "Single-shot, sub-pm-res. wavemeter using topologically enhanced optical activity …" CLEO, JTh5A.9. (2019)
[6] Z. Ma et al., "Light guidance beyond cutoff in optical fibers," in CLEO, paper SF1P.2. (2020)
[7] X. Liu et al., "Suppression of Raman scattering by controlling the angular momentum content of fiber modes," CLEO, FM2P.5. (2020)
[8] F. Poletti and P. Horak. "Description of ultrashort pulse propagation in multimode optical fibers," *JOSA B* **25**, 1645-1654 (2008)

EPJ Web of Conferences **243**, 18001 (2020)
EUROPHOTON 2020

https://doi.org/10.1051/epjconf/202024318001

Temporally resolved studies of thermal effects in high power ZGP OPO pumped by high-repetition Ho:LLF MOPA system

Marcin Piotrowski[1], Manuel-Alessandro Medina[1,2], Martin Schellhorn[1], Anne Hildenbrand-Dhollande[1]

1. French-German Research Institute of Saint-Louis (ISL), 5 rue du Général Cassagnou, 68301 Saint-Louis, France

2. Aix Marseille Univ., CNRS, Centrale Marseille, Institut Fresnel, Marseille, France

Tunable high power laser sources covering mid-IR region (3 – 5 µm) find numerous applications in remote sensing, free space communication or defence countermeasures. Besides high power, such applications require a broad tuning range and a good beam quality. The optical parametric oscillation (OPO) and amplification (OPA) with non-linear zinc germanium phosphide (ZGP) crystals is a main approach to deliver such laser sources [1]. To date, there are only few reports of high average power from ZGP OPOs, often with a compromised beam quality and stability [2-4]. Here, we experimentally investigate methods to enhance a good beam quality, focusing particularly on the aspect of different cavity geometries, lengths, and the influence of a pump repetition rate.

We employed a high-efficient resonantly pumped Q-switched Ho:LLF MOPA 10 kHz system as a pump source for the ZGP OPO. The Ho:LLF pump delivers 68.7 W at 2065 nm in TEM_{00} operation at the repetition rate of 10 kHz with an optical-to-optical efficiency of 61.5 % [5].

Our single stage ZGP OPO generates 38 W of total power in a linear cavity and above 20 W in non-planar ring cavities RISTRA and FIRE [6-8] with $M^2 < 1.5$. Fig. 1 contains the example results for the 130 mm long RISTRA cavity. The left graph shows how the beam quality (M^2) changes with pump pulse energy at the fixed repetition rate of 10 kHz. The right graph shows the thermal effects build-up process. To capture the M^2 factor as a time function we collected beam images at different positions along the beam at 50 Hz frame rate (20 ms time resolution) using pyroelectric array sensor and reconstructed the beam profile in post-processing.

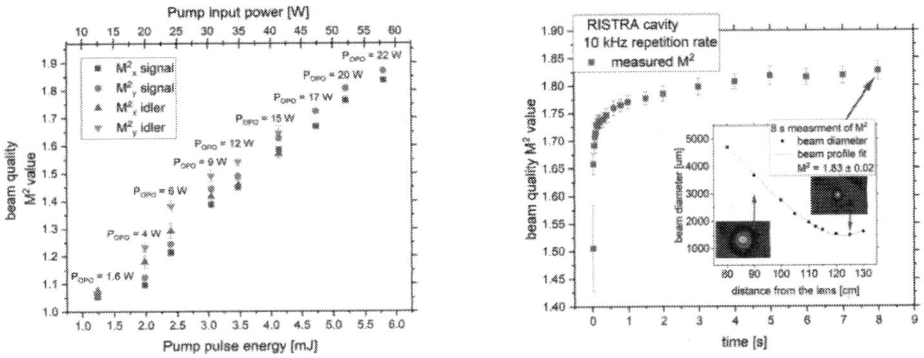

Fig. 1 Left: The OPO output beam M^2 as a function of pump pulse energy at the 10 kHz repetition rate. The labels indicate the total (signal + idler) OPO output power from the non-planar RISTRA cavity. Right: The M^2 time evolution measured with 20 ms resolution at maximum output power of 22 W from ZGP OPO in RISTRA cavity. The inset graph shows the example of a beam profile for the last point in the main graph.

For the first time we provided insight to how thermal effects in a non-linear crystal arise and evolve in time by measuring the M^2 factor of the OPO beam as a time function. We investigated factors affecting the beam quality at high power levels like steady-state and transient thermal lens and methods to compensate for it. Our results can lead to better management of thermal effects and further up scaling of output power from ZGP OPO while maintaining good beam quality.

References

[1] S. Das, "Optical parametric oscillator: status of tunable radiation in mid-IR to IR spectral range based on ZnGeP2 crystal pumped by solid state lasers", Opt. Quantum Electron. **51**, 70 (2019).

[2]. A. Hemming, et al., "99 W mid-IR operation of a ZGP OPO at 25% duty cycle," Opt. Express **21**, 10062–10069 (2013).

[3] C.-P. Qian et. al. "High repetition rate 102 W middle infrared ZnGeP2 master oscillator power amplifier system with thermal lens compensation," Opt. Lett. **44**, 715–718 (2019).

[4] B.-R. Zhao, et. al. "231 W dual end-pumped Ho:YAG MOPA system and its application to a mid-infrared ZGP OPO," Opt. Lett. **43**, 5989–5992 (2018).

[5] M. Schellhorn and G. Spindler, "High-Efficient Resonantly Pumped Q-Switched Ho:LLF MOPA System", in Laser Congress 2019 (ASSL, LAC, LS&C), OSA Technical Digest (Optical Society of America, 2019), paper JTh3A.18.

[6] A. V. Smith and M. S. Bowers, "Image-rotating cavity designs for improved beam quality in nanosecond optical parametric oscillators", J. Opt. Soc. Am. B **18**, 706–713 (2001).

[7] S. Bigotta, G. Stöppler, J. Schöner, M. Schellhorn, and M. Eichhorn, "Novel non-planar ring cavity for enhanced beam quality in high-pulse-energy optical parametric oscillators", Opt. Mater. Express **4**, 411–423 (2014).

[8] M. Schellhorn, G. Spindler, and M. Eichhorn, "Improvement of the beam quality of a high-pulse-energy mid-infrared fractional-image-rotation-enhancement ZnGeP2 optical parametric oscillator", Opt. Lett. **42**, 1185–1188 (2017).

 https://doi.org/10.1051/epjconf/202024318002

Kerr-lens mode locked, synchronously pumped, ultra-broadband breathing pulse optical parametric oscillator

***Jintao Fan*[1,2], *David Zuber*[1,2], *Robin Mevert*[1,2], *Tino Lang*[3], *Thomas Binhammer*[4], *Uwe Morgner*[1,2,5]**

1. Leibniz Universität Hannover, Welfengarten 1, 30167 Hannover, Germany

2. Cluster of Excellence PhoenixD (Photonics, Optics, and Engineering-Innovation Across Disciplines), 30167, Hannover, Germany

3. Deutsches Elektronen-Synchrotron DESY, Notkestraße 85, 22607 Hamburg, Germany

4. neoLASE GmbH, Hollerithallee 17, 30419 Hannover, Germany

5. Laser Zentrum Hannover e.V., Hollerithallee 8, 30419 Hannover, Germany

High repetition rate, ultrabroadband light sources in the near infrared spectral region have gained considerable research interest and emerged as a powerful basis for various applications, such as time-resolved spectroscopy, attoscience, and frequency comb generation [1]. Kerr-lens mode locked (KLM) Ti:sapphire lasers oscillators remain the major workhorse of the research field due to the remarkable spectral bandwidth of this gain material. However, these systems suffer limitation in terms of power scaling mainly owing to unavoidable heat load in the laser crystal. Alternatively, optical parametric oscillators (OPOs) have rapidly developed over the last decade and are a promising candidate to generate high power broadband tunable radiations. However, the pulse forming dynamics in OPOs is more complicated and so far limited the achievable pulse duration to 13 fs [2]. In this work, by introducing a breathing pulse dispersion management scheme, we report the first experimental demonstration of a KLM non-collinear OPO (KLM-NOPO) that emits an ultrabroadband signal spanning from 700 nm to 900 nm at the -10 dB level which would support sub-10 fs pulse durations.

The schematic of the proposed KLM-NOPO is presented in Fig. 1(a). The OPO is pumped by a frequency-doubled home-built Yb-fiber laser. The pump laser has an average power of 7 W at a repetition rate of 50.2 MHz with a FWHM pulse duration 270 fs and a central wavelength 520 nm. The KLM-NOPO is configured as a double foci ring cavity. The 2 mm BBO crystal in one focus serves as the parametric gain medium, while a 1.7 mm Ti:sapphire crystal used as Kerr-medium is placed in another focus. The whole cavity is equipped with double-chirped mirrors (DCMs) that support a flat spectral phase over 600~1200 nm [3]. A pair of BaF_2 wedges is employed to fine adjust the cavity dispersion, and accordingly the overall dispersion of the cavity is close to zero. Note that a strong unbalanced dispersion between two foci is realized in our setup, leading to pulse breathing in one round trip. To this end, a long positive dispersed signal pulse as expected for a good temporal overlap in the BBO together with a short transform limited pulse in the Ti:sapphire is realized. For synchronous pumping, the cavity length is set to match twice the pump lasers repetition rate.

The "magic angle" of 2.5° between the pump beam and the signal beam in non-collinear pointing vector walk-off compensation geometry is selected where ultra-broadband phase matching from 650 to 1200 nm enables broadband operation. By optimizing the position of the Kerr medium, the KLM NOPO can be triggered. As can be seen in Fig. 1(b), this broadband output spectrum covering 670~920 nm provides strong evidence for mode-locked NOPO. This bandwidth results in a transform limited pulse duration of 8.5 fs. Up to 420 mW of output power is measured with the maximum pump power of 7 W. A stable mode-locking performance of our KLM NOPO is indicated by the clean radio-frequency spectrum of the signal (see Fig. 1(c)). In conclusion, the proposed dispersion managed KLM NOPO can generate a stable pulse with an ultra-broadband spectrum capable of handling sub-10 fs pulses. The great potential for power scaling and the flexibility in wavelength of this novel KLM NOPO open the door for the next generation of few-cycle high power laser oscillators.

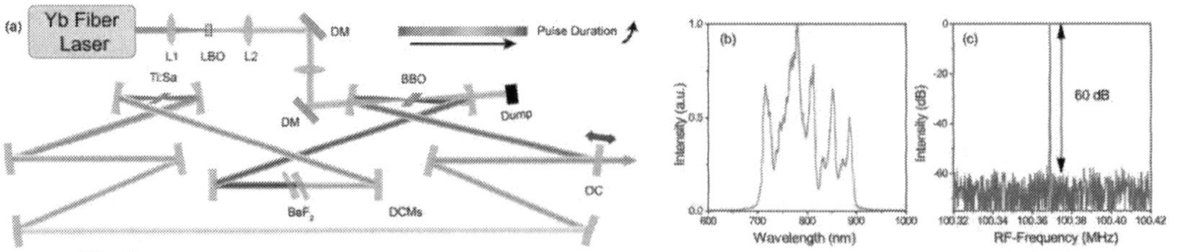

Fig. 1 (a) Schematics illustration of the experimental set-up. L1 and L2: lens; DM: dichroic mirror; DCMs: double-chirped mirror pairs; OC: output coupler. (b) Measured output spectrum (c) Radio-frequency spectrum over a 100 kHz span with 100 Hz resolution.

References

[1] S. A. Diddams, L. Hollberg, and V. Mbele, "Molecular fingerprinting with the resolved modes of a femtosecond laser frequency comb," Nature 445, 627–630 (2007).

[2] G. M. Gale, M. Cavallari, T. J. Driscoll, and F. Hache, "Sub-20-fs tunable pulses in the visible from an 82-MHz optical parametric oscillator," Opt. Lett. **20**, 1562 (1995).

[3] S. Rausch, T. Binhammer, A. Harth, J. Kim, R. Ell, F. X. Kärtner, and U. Morgner, "Controlled waveforms on the single-cycle scale from a femtosecond oscillator," Opt. Express **16**, 9739-9745 (2008)

 https://doi.org/10.1051/epjconf/202024318003

Degenerate Backward wave Optical Parametric Oscillator

Patrick Mutter [1], Andrius Zukauskas [1], Anne-Lise Viotti[2,3], Valdas Pasiskevicius [1] and Carlota Canalias [1]

1. *Department of Applied Phzsics, Royal Institute of Technology, Roslagstullbacken 21, 10691, Stockholm, Sweden*
2. *Deutsches Elektronen-Synchrotron DESY, Notkestraße 85, 22607 Hamburg, Germany*
3. *Department of Physics, Lund University, P.O. Box 118, SE-221 00 Lund, Sweden*

Backward wave optical parametric oscillators (BWOPO) are a special class of nonlinear optical devices with unique properties, due to the fact, that the pump photon is split in two photons propagating in opposite directions. The counter-propagating nature of the interaction automatically establishes positive feedback, without any external cavity, making the system very robust and insensitive to misalignment. Furthermore, the forward wave inherits the phase modulation of the pump, while the backward generated wave is inherently narrow [1]. Although the concept of BWOPO was proposed in the 1960s [2], the fist experimental demonstration had to wait until 2007 [3]. This long delay is ascribed to fact that quasi-phase matched (QPM) structures with sub-um periodicities are required to compensate for the large phase-mismatch between the counter-propagating waves. Thanks to the recent progress in periodic poling based on coercive field engineering, several BWOPO interactions with ps and ns pump pulses have been achieved, demonstrating efficient single pass configuration with conversion efficiencies up to 50% at a mJ output level [4]. BWOPO operating at degeneracy, especially in the telecommunications band, is of special interest due to the possibility to have phase locking inherent to degenerate parametric processes which can be used for generation of biphoton states, a resource for quantum communications and quantum sensing [5, 6]. Efficient BWOPO operation in the spectral range of 1.5 μm -1.6 μm, requires first-order QPM, which can be realized in periodic nonlinear structures with a periodicity of ~400 nm, i.e., close to half of the pump wavelength in vacuum.

In this work, we take advantage of our coercive-field engineering technique [7] to fabricate 435 nm-period in 1 mm thick Rb-doped $KTiOPO_4$ (PPRKTP) crystals. The resulting domain pattern shows excellent uniformity over the entire crystal aperture and a duty cycle close to 50% (see Fig. 1a)) To the best of our knowledge, this poling period marks the shortest QPM period ever fabricated in ferroelectric crystals of comparable aperture. The crystals are used to demonstrate the first realization of a degenerate BWOPO, generating counter-propagating signal and idler in the spectral band of optical communications.

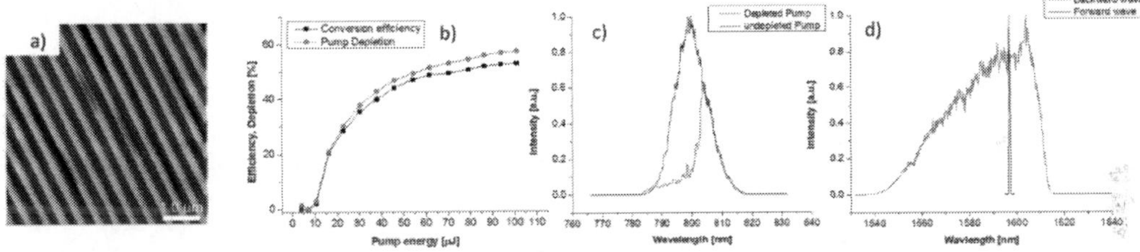

Fig. 1 a) Piezo-force microscope image of one of the polar faces of PPRKTP crystal, b) Combined forward-and backward signal conversion efficiency (black) and pump depletion (red symbols), c) Undepleted Pump (black) and depleted pump (red) d) Forward (red)- and backward wave (black) spectra

The PPRKTP crystal was pumped by a Ti:Sapphire regenerative amplifier delivering 240 ps-long chirped pulses at a central wavelength of 799 nm and a spectral bandwidth of about 10 nm at 1 kHz repetition rate. The laser beam was focused into the crystal to a beam radius of 51 um. The BWOPO process started at a threshold of 265 MW/cm^2. The conversion efficiency reached 53% at a pump energy of 101 μJ (see Fig. 1 b)). Fig c) and d) show the undepleted and depleted pump, the forward wave and backward wave spectra, respectively. As expected, the backward wave was narrowband (0.65 nm) centered at 1597 nm and the forward wave is a spectral replica of the pump. Note that at the degeneracy point (1597 nm) the forward wave show a dip in their spectrum. This is attributed to phase-matched second harmonic generation by the counter-propagating BWOPO signal and idler. This explains the small discrepancy between the pump depletion and the measured BWOPO efficiency. These cascaded interactions bring such functionalities as nonclassical squeezed light generation as well as give opportunity to explore counter-propagating comb generation [8].

References

[1] G. Strömqvist, V. Pasiskevicius, C. Canalias, and C. Montes, "Coherent phase-modulation transfer in counterpropagating parametric down-conversion," *Phys. Rev. A - At. Mol. Opt. Phys.*, vol. 84, no. 2, pp. 1–4, 2011.
[2] S. E. Harris, "Proposed backward wave oscillation in the infrared," *Appl. Phys. Lett.*, vol. 9, no. 3, pp. 114–116, 1966.
[3] C. Canalias and V. Pasiskevicius, "Mirrorless optical parametric oscillator," pp. 459–462, 2007.
[4] R. S. Coetzee, A. Zukauskas, C. Canalias, and V. Pasiskevicius, "Low-threshold, mid-infrared backward-wave parametric oscillator with periodically poled Rb:KTP," *APL Photonics*, vol. 3, no. 7, 2018.
[5] K.-H. Luo *et al.*, "Counter-propagating photon pair generation in a nonlinear waveguide," *Opt. Express*, vol. 28, no. 3, p. 3215, 2020.
[6] C. S. Chuu and S. E. Harris, "Ultra-bright backward wave biphoton source," ," Phys. Rev. A 83, 061803(R) (2011)
[7] Liljestrand, F. Laurell, and C. Canalias, "Periodic poling of Rb-doped KTiOPO4" Opt. Express 24,14682, (2016)
[8] I. Ricciardi, et al "Frequency comb generation in quadratic nonlinear media," Phys. Rev. A, 91, 063839 (2015).

 https://doi.org/10.1051/epjconf/202024318004

Compact picosecond mid-IR PPLN OPO in burst-mode operation

Yudi Wu, Sijing Liang, Qiang Fu, Lin Xu, David J. Richardson
Optoelectronics Research Centre, University of Southampton, Southampton, SO17 1BJ, UK

Short-pulsed mid-infrared (MIR) lasers have many important applications including spectroscopy, material processing and free space communication, where pulse parameters such as a high repetition rate and high peak power can play a critical role in enhancing features e.g. processing speed and signal-to-noise ratio [1]. Optical parametric oscillators (OPOs) are commonly used as MIR laser sources but they require temporal synchronisation between the pump and signal when used to produce short pulses [2]. Hence, picosecond (ps) OPO systems normally operate with a short cavity length for high repetition rate pulses with low peak power, or with a relatively long cavity length for pulses at low repetition rate and high peak power. Here we report a compact short-cavity-length ps OPO that can still provide high-peak-power pulses through burst-mode operation. MIR pulses at a high repetition rate of 1.5 GHz and controllable peak powers up to a maximum of 1.2 kW are demonstrated.

The pump source for the OPO is a 1040-nm gain-switched laser-diode (GSLD) seeded Ytterbium fibre Master Oscillator Power Amplifier (MOPA) system, similar to the MOPA described in reference [3]. The GSLD technique offers highly controllable pulses at a choice of repetition rates and with flexible pulse duration. In this work, a pulse width of 37 ps and repetition rate of 1.5 GHz are selected for the experiment. In the MOPA, an electro-optic modulator (EOM) is employed as a time gate to suppress a variable number of pulses every 1 μs to enable burst-mode operation with a 1-MHz burst repetition rate (Fig. 1a). The MOPA generates 14-W of maximum average output power for all numbers of pulses-per-burst.

The OPO is of a signal-resonant linear cavity design consisting of two concave mirrors with a 40-mm-long periodically poled lithium niobate (PPLN) nonlinear gain medium. Both mirrors are of high reflectivity at a wavelength of 1.5 μm (signal) and ~81% transmission at both 1 μm (pump) and 3 μm (idler). The cavity length was set at just 15 cm to allow synchronous pumping every two signal roundtrips at a 1.5 GHz repetition rate. The PPLN crystal, having 5 channels of grating periods ranging from 29.52 – 31.59 μm, is mounted in an oven that allows temperature tuning from 20 to 200 °C, from which an idler wavelength tunability of 2260 – 3573 nm was realized (Fig. 1b).

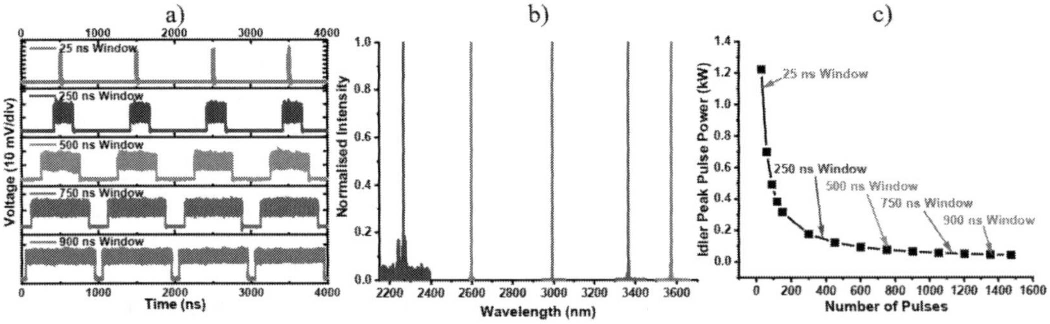

Fig. 1 a) Burst-pulse traces of pump with different time window, b) tunable spectrum of idler, c) idler single-pulse peak power against the number of pulses per burst.

By varying the burst-pulse window with the MOPA's EOM, different numbers of pulses in each burst were generated from the OPO. The average power of the idler output was measured to vary from 1.4 to 2.4 W, dependent upon the window setting. The corresponding peak power of each individual idler pulse of was then calculated and Fig. 1c shows the pulse peak power against the number of pulses for each burst mode. As the number of pulses per burst decreases, the idler pulse peak power increases significantly. With 20 pulses per burst, an idler pulse peak power of 1.2 kW was achieved. This is two orders of magnitude higher than the idler pulse peak power of ~30 W achieved when the OPO was operating with a continuous 1.5 GHz train of pulses. Thus, by using burst-mode operation, a highly compact MIR OPO with high intra-burst repetition rate as well as high peak power was achieved.

References

[1] Y. Su, W. Wang, X. Hu, H. Hu, X. Huang, Y. Wang, J. Si, X. Xie, B. Han, H. Feng, Q. Hao, G. Zhu, T. Duan, and W. Zhao, "10 Gbps DPSK transmission over free-space link in the mid-infrared," Opt. Express **26**, 34515 (2018)

[2] Q. Fu, L. Xu, S. Liang, P. C. Shardlow, D. P. Shardlow, S. Alam, and D. J. Richardson, "High-average-power picosecond mid-infrared OP-GaAs OPO," Opt. Express **28**, 5741 (2020)

[3] H. Chan, S. Alam, L. Xu, J. Bateman, D. J. Richardson, and D. P. Shepherd, "Compact, high-pulse-energy, high-power, picosecond master oscillator power amplifier," Opt. Express **22**, 21938 (2014)

EPJ Web of Conferences **243**, 19001 (2020)
EUROPHOTON 2020
https://doi.org/10.1051/epjconf/202024319001

Overview of company EKSPLA and products offered

G. Kudaba
Ekspla UAB, Savanorių av. 237, Vilnius, LT-02300, Lithuania

The Lithuanian laser manufacturer Ekspla was founded in 1992 as the 11 people spin-off of the laser division from the parent company EKSMA founded in 1983 as a pilot plant at the Institute of Physics.

The first laser made by Ekspla was the picosecond laser PL2140, delivered to the Japanese customer in 1993.

Soon after that, Ekspla gained worldwide recognition as the manufacturer of high performance solid-state Nd:YAG lasers, especially of picosecond and nanosecond time domains, as well as widely tunable parametric generators, SFG, and other nonlinear spectroscopy set-ups. This was confirmed by the "Prism Awards for Photonics Innovation" grant, which was received for the development of the NT200 tunable nanosecond tunable system in the 2011 year.

Through continuous growth, Ekspla has developed to 120 employees company with more than 18 Ph.D.'s. EKSPLA actively participates in local and EU projects, investing more than 1.5 mln. EUR for research and development annually. EKSPLA is widely represented in the world through its broadly developed distribution network. With more than 2500 lasers sold, EKSPLA is well known among top universities and research institutions, like NASA, CERN, Max Planck Institute, Massachusetts Institute of Technology, JAXA, University of Cambridge and others.

EKSPA has been continuously growing its product portfolio and now is offering standard and specially designed lasers for the science and industry applications. Among others, it is worth to mention such installations as the most powerful 5.0 TW @ 880 nm, 1 kHz, 9 fs OPCPA system in ELI-ALPS, Hungary together with Light Conversion; the most powerful laser in the world 10 PW Extreme Light Infrastructure in the Czech Republic together with US company National Energetics; 160 J nanosecond Nd: phosphate glass laser in South Korea.

One of the recent EKSPLA developments is the NL740 series laser with highly stabile output parameters:

- Narrow bandwidth, stable, true SLM pulses
- Excellent pulse energy (typically 0.1 % StDev @ 1064 nm) and pulse duration stability
- Excellent spatial mode stability
- Excellent output power stability (typically < ±0.5 % peak-to-peak)
- 3 – 10 ns tunable pulse duration
- Up to 100 mJ output energy
- Up to 100 Hz repetition rate
- 1064, 532 nm or 355 nm output wavelength
- Reliable 24/7 operation

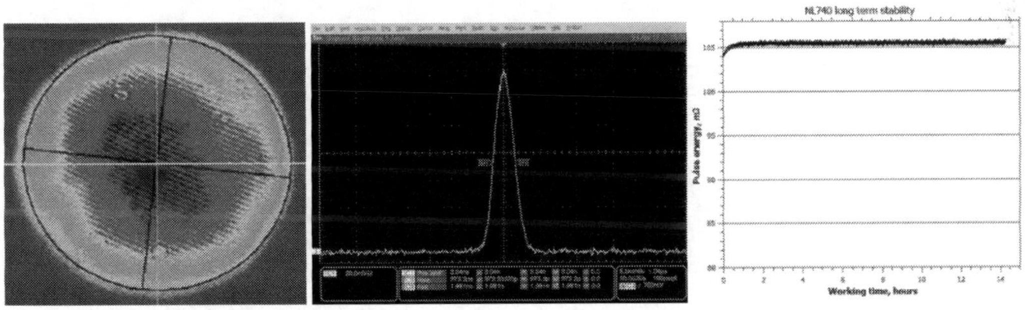

Fig. 1 The typical near field beam profile, pulse shape and long-term energy stability of NL740 series laser.

NL740 series laser (model NL940-100-SH/TH) was successfully installed for the Laser-Induced Damage Threshold (LIDT) application [1]. The company LIDARIS used it for investigation of UV optics reliability in the joint project "ESsential PREparation Steps for Qualification Longevity of Space Optics" (ESPRESSO, PECTS call, Contract Nr. 4000115898) with the European Space Agency (ESA/ESTEC).

References

[1] A. Melninkaitis, G. Batavičiūtė, C. Heeseb, M. Ščiuka, and L. Smalakys, "Towards qualification longevity of high power space optics," Proceedings of the SPIE, **11180**, id. 1118085 12 pp. (2019).

Overview on iXblue Photonics and focus on ModBox-FrontEnd Solution

Hervé Gouraud
iXblue Photonics – 3, rue Sophie Germain, F-25000 Besançon, France

iXblue is a global high-tech company specializing in the design and manufacturing of advanced autonomous, marine and photonics technologies. iXblue Photonics is a division of iXblue, it helps photonics engineers all around the world to get the most out of the light by providing high performance, innovative and reliable photonic solutions. iXblue Photonics solutions are dedicated to diverse industrial applications such as Telecom, Sensing, Defense, Space, Fiber Laser, Harsh Environment.

The Modulation Solutions Division from iXblue Photonics is specialized in the design and manufacture from low to 70 GHz high frequency $LiNbO_3$ of optical fibered modulators, and matching components such as microwave amplifiers and electronics.

The Specialty Fibers Division from iXblue Photonics develops and manufactures a wide range of specialty optical fibers, fiber optic components and subassemblies. iXblue specialty fibers and components are produced in environmentally controlled production areas to ensure high proof strength, tight tolerances and a high-quality level.

By mastering all the building blocks of an efficient modulation system, iXblue is capable to offer complex optoelectronic modulation solutions based on proprietary designs such as the ModBox-FrontEnd. The ModBox-FrontEnd is an optical sub-ns pulse shaping laser: it is a carefully tuned turnkey solution dedicated to a specific High Energy or High-Power class laser. The ModBox-FrontEnd is used to seed high power amplifier chains to produce a master laser or an OPCPA pump laser. Applications range from scientific research and academic activities to industrial market needs. The ModBox-FrontEnd operates in many different fields such as particle acceleration (GSI), inertial confinement fusion (LLNL), radiation therapy, X-ray generation, for sensor test applications (Arété Associates), laser-driven shock compression for laser peening (Hilase) or dynamic compression (ESRF). iXblue offers dedicated industrial grade Fibre Front End Seed Sources for all industries including R&D, Defense or production.

"The ability to shape the pulse temporally is of critical importance": the ModBox-FrontEnd will challenge the state of the art of temporal pulse shaping performance:

- Operating wavelength 1030 nm, 1053 nm, 1064 nm
- Output pulse energy 1 nJ per ns
- Very high pulse contrast up to 60 dB
- Excellent output power stability (typically < ±1 % peak-to-peak)
- 125 ps – 100 ns tunable pulse duration with pulse shaping capability
- Up to 100 kHz repetition rate
- Reliable 24/7 operation

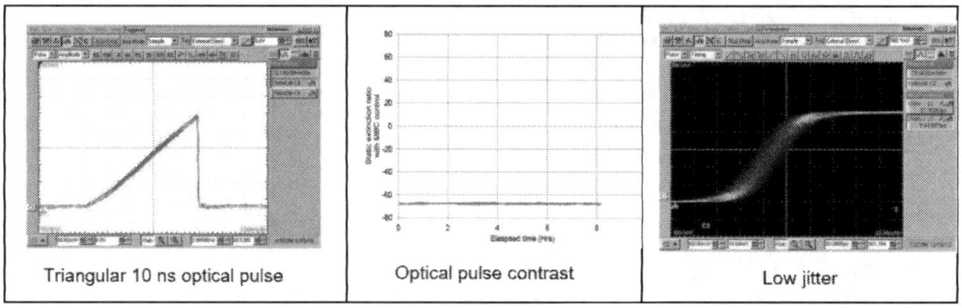

Fig. 1: Pulse Shaping and ModBox optical performances

High-energy spatiotemporal solitons in GRIN fiber

F. Mangini[1], M. Zitelli[2], M. Ferraro[2], D.S. Kharenko[3], A. Niang[1], A. Tonello[4], V. Couderc[4], S. Wabnitz[2,3]

1. *Department of Information Engineering (DII), University of Brescia, Via Branze 38, 25123 Brescia, Italy.*
2. *Department of Information Engineering, Electronics and Telecommunications (DIET), Sapienza University of Rome, Via Eudossiana 18, 00184 Rome, Italy.*
3. *Novosibirsk State University, Pirogova 1, Novosibirsk 630090, Russia.*
4. *Université de Limoges, XLIM, UMR CNRS 7252, 123 Avenue A. Thomas, 87060 Limoges, France*

In the last decade, the study of the propagation of ultrashort pulses in multimode (MM) fibers has seen a renewed interest, thanks to the potential that this technology can offer for guided optical communications [1–3]. Despite numerous studies in the literature, several aspects are still to be fully understood. In this work, we consider the high-energy regime, up to reaching the fiber damage threshold, of spatiotemporal soliton generation in MM fibers. In particular, we study the propagation of femtosecond pulses in a MM fiber with a graded-index (GRIN) core profile. Intrapulse Raman scattering leads to the fission of the initial pulse into many MM solitons, which undergo Raman self-frequency shift (SSFS) [4].

In the present study, we reveal the presence of a new type of nonlinear propagation regime in MM fibers, where stable spatiotemporal solitons are created by the fission of the initial pulse. These solitons have different amplitudes and wavelengths, but nearly equal time duration. To study the generation and fission of high-energy MM solitons in the anomalous dispersion regime, we used a hybrid optical parametric amplifier (OPA) generating pulses with hyperbolic secant temporal profile of 130-70 fs duration at 1550 nm, and repetition rate of 33 kHz. The input pulse energy was varied between 0.1 nJ and 500 nJ by a variable attenuator. The laser beam was coupled with a 30 cm long multimode standard 50/125 GRIN fiber with relative index difference $\Delta = 0.0103$ (cladding refractive index is 1.457) and chromatic dispersion $\beta_2 = -22$ ps^2/km by a 50 mm lens producing a 30 µm beam waist at the fiber input. Numerical simulations were also conducted to reproduce the phenomenon of fission using an exact 3D+1 vector model that includes second, third, and fourth-order dispersion, Kerr, and Raman nonlinearities. We neglected the contribution of polarization mode dispersion, since it is negligible for the fiber lengths involved in our experiments. We included a phenomenological two-photon absorption (TPA) term, to take into account the presence of nonlinear losses (TPA coefficient is taken to be $\alpha_2 = 10^{-16}$ m/W by fitting the experimental SSFS values). Random phase noise was added to the input field, to describe the generation of intensity speckles in the output transverse field, and to seed the generated dispersive waves.

In Fig.1.a the output spectrum is shown for different values of the input pulse energy. It shows the creation of several Raman solitons. In particular, one can observe how the Raman-induced SSFS tends to saturate for energies higher than 200 nJ. Correspondingly, the total output energy is clamped, as shown in Fig. 1.b : we revealed that this is due to the presence of high (up to 80% transmission drop) nonlinear loss in the first few cm of MM fiber. We ascribe the nonlinear loss to multiphoton absorption by defects of the Germanium doped fiber, which is enhanced at the points of peak intensity, owing to beam self-imaging. Therefore, two distinct propagation regimes exist: in the first, only weak linear losses are present ; in the second, the output energy remains clamped to a nearly constant value. From Fig.1.b, we can see that 100-150 nJ is the input energy boundary between these two regimes. Fig. 1.c shows that the soliton pulsewidth remains nearly constant (50-60 fs) as the soliton energy grows larger than about 10-15 nJ. Remarkably, in the nonlinear loss regime nearly all of the transmitted energy is funneled into high-energy spatiotemporal soliton pulses with a bell-shaped, high-quality transverse beam profile.

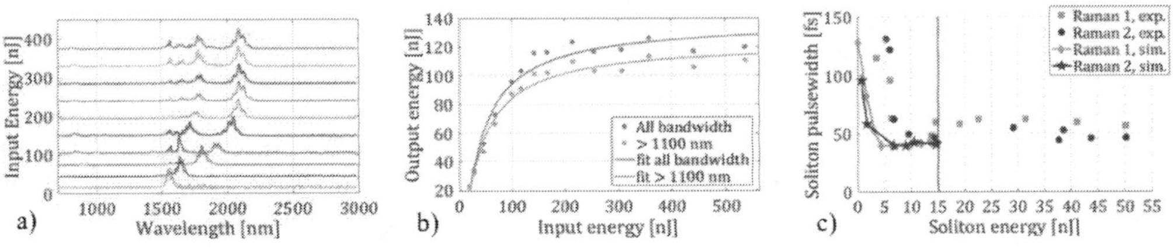

Fig. 1 a) Output spectrum evolution at different output energy (linear scale); b) experimental output energy vs. input energy, in the overall bandwidth and for wavelength above 1100 nm; c) Soliton pulsewidth vs. soliton energy.

This work was supported by the European Research Council (grant No. 740355), and by the Russian Ministry of Science and Education, (grant No. 14.Y26.31.0017).

References

[1] W. H. Renninger et al., "Optical solitons in graded-index multimode fibres," Nat. Commun. **4**, 1719 (2013).
[2] L. G. Wright, et al., "Spatiotemporal dynamics of multimode optical solitons," Opt. Express **23**, 3492 (2015).
[3] K. Krupa, et al., "Multimode nonlinear fiber optics, a spatiotemporal avenue," APL Photonics **11**, 110901 (2019).
[4] A.S. Ashan and G.P. Agrawal, "Spatio-temporal enhancement of Raman-induced frequency shift in graded-index multimode fibers," Opt. Lett. **44**, 2637 (2019).

3D printed Er-doped silica fibre by Direct Ink Writing

Angeles L. Camacho Rosales*[1], Martín Núñez-Velázquez[1], Jayanta K. Sahu[1]
[1]*Optoelectronic Research Centre, University of Southampton, University Road Southampton SO17 1BJ*

Over the past decades, many new types of optical fibres have been developed to cope with the ever-increasing demand for bandwidth in optical communications, and to extend the use of optical fibers in other fields. Some of the features of these optical fibres are complex geometries and multiples materials such as photonics crystal fibres (PCFs), anti-resonant fibers (ARFs) and multicore fibres (MCFs), among others. The current fabrication methods often require a lengthy and complex preform fabrication processing that compromise the mechanical integrity of the preform, Additive manufacturing processes have been explored to tackle the current drawbacks of optical fibres fabrication. In recent years, direct ink writing (DIW)[1], Stereolithography (SLA)[2], digital light processing (DPL)[3], and laser powder bed fusion (LPBF)[4] have been used to 3D print glasses. Moreover, 3D printed silica optical fibres with complex structures[5] and multiple materials[6] were reported using laser printing method[5,6]. In this work, we present the fabrication of a 3D printed Er-doped silica optical fibre by DIW method.

The 3D printed preform was fabricated using a composite mixture of 12wt% hydrophobic fumed SiO_2; 2wt% Polydimethylsiloxane (PDMS); 85wt% of tetraethylene glycol dimethyl ether (Tetraglyme), and 1wt% of erbium chloride. A 3D printer, Ultimaker+2, was used to deposit the composite mixture in uniform layers of 500μm thickness and a constant printing speed of 40mm/s at room temperature. A printed green body was produced, which is subsequently heat treated to a transparent glass as described in Fig 1. The 3D printed Er-doped preform was placed inside a fluorinated tube that was used to provide the refractive index contrast to the fibre. A multimode Er-doped optical fibre was drawn with an OD of 100μm and core diameter of ~40μm. The compositional analysis using energy-dispersive X-ray spectroscopy (EDX) on the fibre confirmed the incorporation of erbium ions into the silica glass. A white light source (WLS) was used to evaluate the Er^{3+} absorption in the core region. The produced 3D printed optical fibre presents an absorption of 62.98dB/m at 980nm and 151.49dB/m at 1535nm, as shown in Fig 2. More optical characterisation results will be presented at the conference.

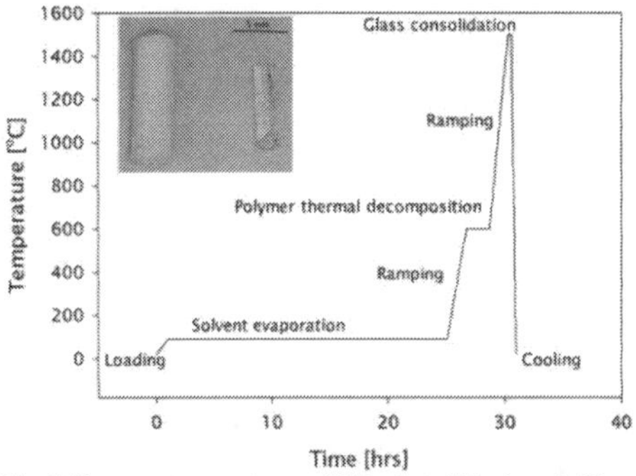

Fig. 1 Heat treatment stages of 3D printed Er-doped silica green body. (inset, 3D printed preform is shown after drying and after consolidation into transparent glass).

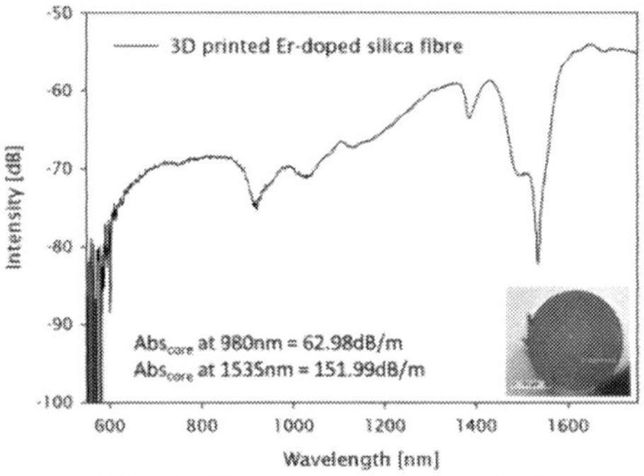

Fig. 2 Absorption spectrum of 3D printed Er-doped silica optical fibre by DIW.

In summary, we have demonstrated the DIW process for fabricating RE-doped 3D printed silica optical preforms. The reported optical fibre presents additional absorption peaks at wavelengths ~1380nm and ~1030nm, which are likely to be the contribution of residual OH, PDMS and impurities from the starting materials. A refinement of the fabrication process is needed for elimination of residual polymer and OH.

The authors would like to acknowledge the contributions from the UK Engineering and Physical Sciences Research Council award: EP/M020916/1; and CONACyT-Mexico awards: CVU607008 and CVU377962. All data supporting this study is available at https://doi.org/10.5258/SOTON/D1418.

References

[1] D. T. Nguyen, et.al. "3D-Printed Transparent Glass", Adv. Mat. 29, 5. 201701181 (2017).
[2] F. Kotz, et.al. "Three-dimensional printing of transparent fused silica glass" Nature 544, 337, 22061 (2017).
[3] E. Shukrun, et.al. "3D-Printed Organic–Ceramic Complex Hybrid Structures with High Silica Content". Adv. Science 5, 1800061, (2018).
[4] K.C. Datsiou, et.al. "Additive manufacturing of glass with laser powder bed fusion". J AM CERAM SOC 102, 4410-4414, 16440 (2019).
[5] A.L. Camacho-Rosales, et.al. "Development of 3-D Printed Silica Preforms" in CLEO /Europe-EQEC 2019 Ed 2019.
[6] A.L. Camacho-Rosales, et. al. "Optical fibers fabricated from 3D printed silica preforms". in PW (SPIE, 2020) Vol. 11271 PWL.

Hybridizing Multi-pass and Multi-plate Bulk Compression

M. Seidel,[1] P. Balla,[1] T. Binhammer,[2] M. Frede,[2] G. Arisholm,[3] L. Winkelmann,[1] I. Hartl,[1] and C.M. Heyl[1,4]

1. Deutsches Elektronen-Synchrotron DESY, Notkestraße 85, 22607 Hamburg, Germany
2. neoLASE, Hollerithallee 17, 30419 Hannover, Germany
3. FFI (Norwegian Defence Research Establishment), P. O. Box 25, NO-2027 Kjeller, Norway
4. Helmholtz-Institute Jena, Fröbelstieg 3, 07743 Jena, Germany

With the advent of ultrafast Yb-ion based disk, slab and fibre lasers, nonlinear pulse compression methods became indispensable for the generation of high average power sub-100 fs pulses. In particular, spectral broadening in Herriott-type *multi-pass* cells (MPC) [1] has been established as a novel tool for pulse compression in the 5 – 100 MW peak power range where neither solid-core fibre nor hollow-core capillaries work efficiently. The operation in the critical self-focusing regime becomes feasible through repetitive refocusing of the cell mirrors. Analogously, the *multi-plate* approach [2], which has mainly been used for broadband continuum generation, relies on the nonlinear refocusing of the Kerr media [3]. By introducing the hybridization of both methods, we demonstrate compression factors up to 14. To the best of our knowledge this exceeds all single-stage compression factors achieved with bulk-based spectral broadening to date (Fig. 1e).

In our experiments, we used a home-built fibre-laser front-end and a neoLASE Yb:YAG amplifier [4]. The source delivered 10 Hz bursts of up to 800 pulses with 100 μJ energy, 880 fs duration and 1 MHz repetition rate. Our MPC consisted of dielectric focusing mirrors with 200 mm radius of curvature. Inserting a single 6 mm thick silica plate resulted in broadening to a Fourier limit of about 100 fs and the characteristic phase jump in the retrieved FROG trace known from bulk compression [3]. However, when using three 1 mm thick plates with about 3 cm mutual distance, the phase jump was not observed anymore and clean pulse compression was achieved (Fig. 1 b). With a 380 mm long cell and 85 μJ pulse energy, we compressed the pulses to 65 fs by means of chirped mirrors (Fig. 1 c). The full laser output was sent through a 350 mm long MPC with 85 % transmission. The pulses were compressed to 75 fs with a grating pair. The M^2 factor was < 1.4 and the spectrum was stable over a measurement period of 15 hours. A further increase of input energy and broadening factor for the 380 mm cell is possible if the mode-matching of the input beam is adjusted to the nonlinear self-lensing of the Kerr media. Simulations with the SISYFOS package [3] show that, contrary to linear (or low power) mode-matching, the peak intensity in the nonlinear media is kept well below the damage threshold. Moreover, compression factors exceeding 20 are predicted if 6 instead of 3 plates are placed into the MPC. Furthermore, the nonlinear effects become saturated, yielding an additional increase of the output spectrum stability (Fig. 1d).

In conclusion, we have shown that hybridizing multi-plate and multi-pass bulk broadening techniques results in a compact pulse compression setup capable of generating clean, nearly Fourier-limited sub-100 fs pulses. We have experimentally demonstrated compression factors of more than 10 and simulated factors of more than 20. These are expected to be experimentally confirmed soon. With that, the simple-to-implement and robust bulk broadening method becomes also increasingly competitive to gas-filled MPCs [5, 6].

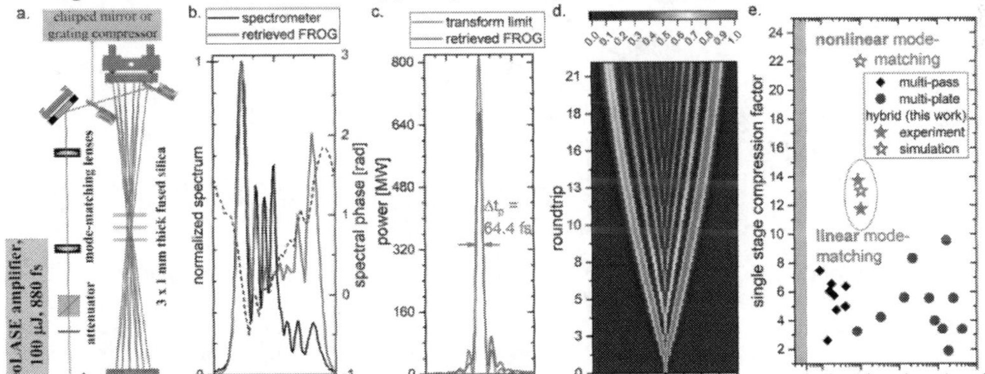

Fig. 1 **a.** Experimental setup. **b.** Retrieved FROG spectrum in comparison to grating spectrometer spectrum after 23 roundtrips in the 380 mm cavity. **c.** Corresponding retrieved pulse in comparison to the Fourier transformed spectrum. **d.** Simulation of spectral broadening with 6 plates in the MPC and nonlinear mode-matching. **e.** Reported single stage compression factors of bulk multi-pass and multi-plate experiments in comparison to the results presented here. Encircled are the experimental 75 fs and 65 fs compression results (solid red stars) and the benchmarking simulation (hollow star). The hollow star at the top predicts 40 fs pulses for 6 plates if mode-matching accounts for self-focusing.

References

[1] J. Schulte et al., Opt. Lett. **41,** 4511 (2016).
[2] C.-H. Lu et al., Optica **1,** 400 (2014).
[3] M. Seidel et al., Opt. Exp. **24**, 9412 (2016).
[4] O. Akcaalan et al., submitted to Europhoton 2020.
[5] M. Kaumanns et al., Opt. Lett **43,** 5877 (2018).
[6] P. Balla et al., Opt. Lett. **45,** 2572 (2020).

Post-compression of high average power picosecond pulses for few cycle generation and FEL pump-probe experiments

Anne-Lise Viotti[1,2], Skirmantas Alisauskas[1], Prannay Balla[1,3], Ammar Bin Wahid[1], Ivan Sytcevich[2], Chen Guo[2], Laura Silletti[1], Andrea Cartella[4], Hamed Tavakol[1], Uwe Grosse-Wortmann[1], Arthur Schönberg[1,3], Marcus Seidel[1], Bastian Manschwetus[1], Tino Lang[1], Andrea Trabattoni[1], Francesca Calegari[1,4,5], Arnaud Couairon[6], Anne L'Huillier[2], Cord L. Arnold[2], Ingmar Hartl[1] and Christoph M. Heyl[1,3]

[1]*Deutsches Elektronen-Synchrotron DESY, Notkestraße 85, 22607 Hamburg, Germany*
[2]*Department of Physics, Lund University, P.O Box 118, SE-221 00 Lund, Sweden*
[3]*Helmholtz-Institute Jena, Fröbelstieg 3, 07743 Jena, Germany*
[4]*The Hamburg Centre for Ultrafast Imaging, Universität Hamburg, 149 Luruper Chaussee, Hamburg 22761, Germany*
[5]*Institut für Experimentalphysik, Universität Hamburg, Luruper Chaussee 149, 22761 Hamburg, Germany*
[6]*Centre de Physique Théorique, CNRS, Ecole Polytechnique, Institut Polytechnique de Paris, F-91128 Palaiseau, France*

Ultrafast Ytterbium-based (Yb) amplifiers can be power-scaled into the kilowatt (kW) regime, offering great potential for applications where repetition rate scaling of ultrafast lasers beyond the average power limits of Ti:sapphire technology is needed. Examples are future laser-driven accelerators and pump-probe experiments at superconducting X-ray free electron lasers (FEL). However, the gain bandwidth of Yb media limits the output pulse durations of such amplifiers to a few 100s of fs to 1 ps. External spectral broadening concepts can be employed for pulse post-compression, offering the possibility to combine ultrashort pulse durations, high repetition rates and milli-joule (mJ) level pulse energies. In particular, a recently developed post-compression scheme based on spectral broadening in gas-filled multi-pass cells (MPC), shows good prospects for peak- and average-power scaling [1].

Here, we demonstrate a compression factor of about 40 in a single pulse compression stage, compressing 1.2 ps, 2 mJ pulses centered at 1030 nm to 32 fs at 200 W in-burst average power. A second MPC stage is used to reach pulse durations down to 13 fs, thus opening a route towards high average power few-cycle sources driven by commercial high-power ps lasers [2]. The experimental setup is shown in Fig. 1 (a). It consists of two compression stages containing mode-matching units, gas-filled MPCs and dispersion compensating mirrors (DCMs). The cells are operated close to the stability edge, at lengths close to 2 m, with 44 passes for the first cell and 12 passes for the second cell. The compressed output pulses are characterized by SHG-FROG and dispersion scan, respectively. The results are shown in Fig.1 (b).

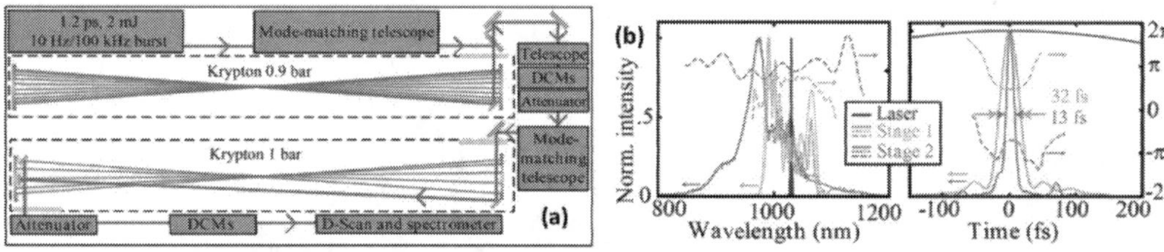

Fig. 1 (a) Experimental setup consisting of two gas-filled MPCs, mode-matching optics and DCM compressors. (b) Reconstructed spectral and temporal intensity (solid) and phase (dashed) profiles together with the corresponding measured spectra after both MPCs (dotted) and at the laser output (dark blue). SHG-FROG data (stage 1) are shown in light blue and dispersion scan data (stage 2) are shown in green.

The throughput of the first MPC is 85 % (80 % including the DCM compressor). The throughput of the second MPC is 46 %, limited mainly by losses at the MPC silver mirrors, yielding 0.37 mJ pulses at the output.

These results are obtained employing a commercial Yb;YAG Innoslab amplifier system, which is currently used as pump for the OPCPA pump-probe laser system at the FLASH FEL facility. With a relatively simple add-on, the here demonstrated post-compression unit can complement the parameter range of the FLASH pump-probe laser system available for laser-FEL pump-probe spectroscopy [3]. In particular, long-term stability measurements are carried out with a single MPC in preparation for first FEL user experiments. Over 36 h, MPC output pulse energy fluctuations are below 4% RMS (2% at the laser output) for compressed output pulses of about 55 fs duration and a standard deviation of 2.8 fs. These results, together with a well-preserved beam quality through the MPC (M^2 of 1.3x1.2 vs. M^2 of 1.2x1.1 for the laser output), show that such a post-compression system is suitable for various ultrafast measurements including applications which demand long-term stable operation.

References

[1] J. Schulte *et al.*, "Nonlinear pulse compression in a multi-pass cell", Opt. Lett. **41**, 4511 (2016).
[2] P. Balla et al., "Postcompression of picosecond pulses into the few-cycle regime", Opt. Lett. **45**, 2572 (2020).
[3] J. Rossbach *et al.*, "10 years of pioneering X-ray science at the Free-Electron Laser FLASH at DESY," Phys. Rep. **808**, 1 (2019).

 https://doi.org/10.1051/epjconf/202024321003

Overcoming gas ionization limits with divided-pulse nonlinear compression

G. W. Jenkins, C. Feng, J. Bromage

Laboratory for Laser Energetics, University of Rochester, 250 East River Road, Rochester, NY 14623-1299, USA

Gas-filled hollow-core fibers (HCFs) have been successfully used for spectrally broadening high-average-power Yb systems to ~10-fs pulse durations for a number of applications. However, pulse-energy scaling in an HCF is limited by self-focusing and ionization [1]. Our simulations show that divided-pulse nonlinear compression (DPNLC), shown schematically in Fig. 1(a), can overcome those limits and improve the spectral broadening of high-energy pulses.

The self-focusing in an HCF can be controlled by reducing the nonlinear index of the gas as the peak power of the pulse is increased [2,3]. The gas pressure (p) is a convenient variable to tune the nonlinear index and should be set below Tempea and Brabec's threshold of 9% peak power in the second-order mode. In our simulations, we set the gas pressure to half this value and vary the fiber radius to vary the strength of plasma effects. We simulate a 10-mJ, 1-ps (FWHM) pulse centered at a 1030-nm wavelength, propagating through the HCF with a model adapted from Horak and Poletti [4]. Figure 1 shows that the onset of ionization at a fiber diameter of about 550 μm leads to significant energy losses in the fiber and a reduction in the peak power achievable.

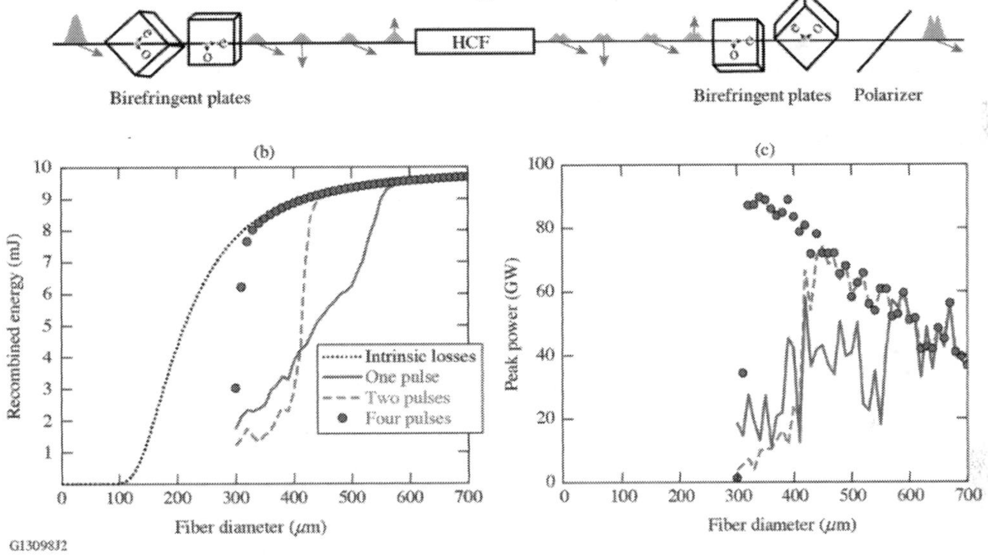

Fig. 1 (a) Schematic of DPNLC. (b) Simulated output energy after HCF and divided pulse recombination stages for a range of fiber diameters. Onset of ionization is clearly visible around 550 μm, 400 μm, and 300 μm for one, two, and four pulses, respectively, and results in large energy losses. (c) Simulated peak power after recombination and compression with up to second-order phase.

In DPNLC, a high-energy pulse is divided into multiple low-energy pulses, which are spectrally broadened in the fiber (note that the gas pressure is re-optimized for the low-energy pulses), and then they are recombined to form a high-energy, spectrally broadened pulse [5]. Figure 1 displays our simulations of the recombined pulse parameters if two 5-mJ, or four 2.5-mJ pulses are spectrally broadened instead of one 10-mJ pulse. The ionization limit is shifted to smaller fiber diameters, leading to higher energy throughputs, larger spectral broadening factors, and higher peak powers after the pulses are recombined.

This material is based upon work supported by the Department of Energy National Nuclear Security Administration under Award Number DE-NA0003856, the University of Rochester, and the New York State Energy Research and Development Authority. The support of DOE does not constitute an endorsement by DOE of the views expressed in this abstract.

References

[1] C. Vozzi, M. Nisoli, G. Sansone, S. Stagira, and S. De Silvestri, "Optimal Spectral Broadening in Hollow-Fiber Compressor Systems," Appl. Phys. B **80**, 285 (2005).
[2] G. Tempea and T. Brabec, "Theory of Self-Focusing in a Hollow Waveguide," Opt. Lett. **23**, 762 (1998).
[3] N. Milosevic, G. Tempea, and T. Brabec, "Optical Pulse Compression: Bulk Media Versus Hollow Waveguides," Opt. Lett. **25**, 672 (2000).
[4] P. Horak and F. Poletti, "Multimode Nonlinear Fibre Optics: Theory and Applications," in *Recent Progress in Optical Fiber Research*, edited by M. Yasin (InTechOpen, Rijeka, Croatia, 2012), pp. 1–24.
[5] H. Jacqmin *et al.*, "Passive coherent combining of CEP-stable few-cycle pulses from a temporally divided hollow fiber compressor," *Optics Letters*, vol. 40, no. 5, pp. 709-712, Mar 2015.

 https://doi.org/10.1051/epjconf/202024322001

Pulsed Single-Frequency Polarization-Maintaining Holmium Laser at 2050 & 2090 nm

D. Lorenz[1,2]***, C. Romano***[1]***, M. Eichhorn***[1,2]***, C. Kieleck***[1]

1. Fraunhofer IOSB (Institute of Optronics, System Technologies and Image Exploitation), Gutleuthausstraße 1, 76275 Ettlingen, Germany

2. Karlsruhe Institute of Technology, Institute of Control Systems (IRS), Fritz-Haber-Weg 1, 76131 Karlsruhe, Germany

We report on the development of a nanosecond pulsed multistage fiber master oscillator power amplifier (MOPA) targeted as a pump source for mid-infrared conversion; long wavelength (≥ 2 µm) pulsed lasers are needed to mitigate losses in the nonlinear crystals [1]. Mid-infrared sources can be used for sensing or medical applications. N. Simakov et al. have shown that Holmium-doped fiber amplifiers (HDFA) are a great solution to amplify signals above 2 µm, with an adjustable bandwidth depending on the pump wavelength [2].

Our MOPA setup consists of a semiconductor laser centered at 2050 nm or 2090 nm as seed (linewidth < 13 MHz), followed by a two-stage HDFA build with polarization-maintaining fibers only. The Amplifier configuration was optimized with a simulation tool. The first stage is operated in CW regime, pumped at 1993 nm. Pulses are generated by external modulation through an acousto-optic modulator (AOM) placed at the output of the first stage and are then amplified by the second stage pumped at 1950 nm.

Figure 1 (a) shows the output power of the first stage versus the input pump power at 1993 nm. For the first stage at 2050 nm we measure a slope efficiency $\eta = 63.2$ % (resp. 59.9 % at 2090 nm) and a maximum output power of 902 mW (resp. 920 mW) for a input power $P_{in} = 9.8$ mW (resp. $P_{in} = 3$ mW), which corresponds to a gain of 19.6 dB (resp. 24.8 dB).

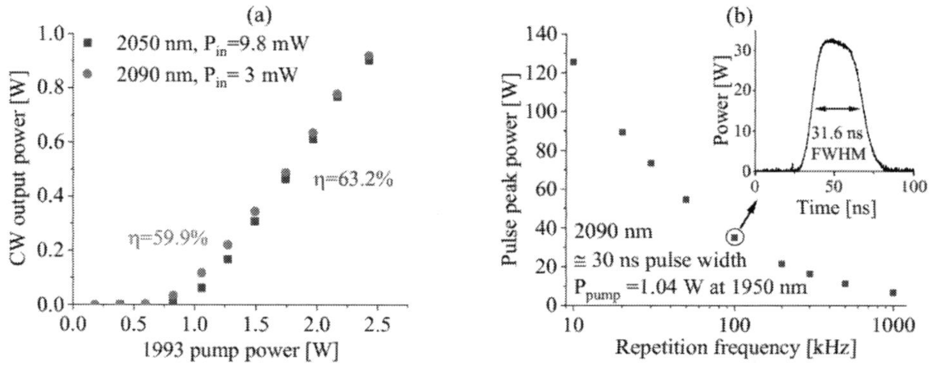

Fig. 1 (a) First stage output power at 2050 nm and 2090 nm versus input pump power at 1993 nm. (b) Second stage peak power for a ≅ 30 ns pulse at 2090 nm versus the repetition frequency.

We investigate the amplification of ≅ 30 ns rectangular pulses from 10 kHz to 1 MHz. The output pulse peak power versus the repetition frequency is shown in figure 1 (b). At a repetition rate of 10 kHz a maximum pulse peak power of 125 W is measured. The inset shows the output pulse shape at 100 kHz repetition rate, the pulse distortion is caused by gain saturation. For repetition rates of 10 kHz up to 100 kHz the pulse peak power out of the second stage was limited by the onset of stimulated Brillouin scattering. Meanwhile no parasitic lasing of the amplified spontaneous emission in between pulses is observed. We present a comparison of the two-stage MOPA performance for a second stage co- or counter-pumped at 1950 nm with the goal of reaching the highest peak power at both signal wavelengths.

References

[1] P. G. Schunemann et al., "Advances in nonlinear optical crystals for mid-infrared coherent sources," J. Opt. Soc. Am. B 33, D36-D43 (2016)

[2] N, Simakov et al., "High gain holmium-doped fibre amplifiers," Opt. Express 24, 13946-13956 (2016)

AUTHOR INDEX

AUTHOR INDEX

AUTHOR INDEX

EDP Sciences – Web of Conferences
17, Avenue du Hoggar
Parc d'Activité de Courtabouf
BP 112
F-91944 Les Ulis Cedex A
France

ISBN 978-1-7138-1921-9

9 781713 819219

CURRAN ASSOCIATES INC.
proceedings
.com

International Conference on Novel Functional Materials (ICNFM2018)

MATEC Web of Conferences Volume 238 (2018)

Anshan, China
16 - 18 September 2018

edp sciences